UNDERSTANDING ETERNITY

Published in 1993 by Psychotechnic Publications, 5 Haig Lane,
Church Crookham, Fleet, Hants., GU13 0UN.

ISBN 1 – 870046 – 09 – 9

Typeset and printed by Mulberry Books, 62bis rue de Vic,
62100 CALAIS, France.

UNDERSTANDING ETERNITY

by Keith Wakelam

PSYCHOTECHNIC PUBLICATIONS

Acknowledgements

To Dr. Jocelyn Bell-Burnell, my patient tutor and Dr. G.A. Hocking, Imperial College, who suggested revisions of the text.

CONTENTS

FOREWORD

The idea of an eternity outside time may seem to belong to the realms of metaphysics, yet science today recognises an all-pervading system of waves called a quantum field which does not obey classical mechanics and instead relies upon chance expectation. We may believe, like Einstein, that 'God does not play dice with the universe', but is there a rational evolutionary principle we can find, like that of Darwin, to explain how the vast scheme of things came into being.

The survival of the fittest cannot be relied upon as a evolutionary principle in cosmology, because it only works when the stage has been set and the actors called upon to play their parts. A law of survival can only operate between a creature and a hostile environment. There seem to be no preferential characteristics to prolong the life of particles, or cause them to organise themselves.

The only law which has been seen to be operating through a universe which was random from the beginning is the one attributed to Murphy, which tends to disorder and chaos.

Given, however, that our material world only occupies a minute instant of time within the boundless duration of eternity, there is scope for theorising the existence of forces and structures which shape our destiny and cause the universe to develop in an ordered fashion. Chaos need not hold sway.

In this study we may have to go back to the beginning and look again at some very basic assumptions. I was once told that a scientific axiom is something that cannot be proved, and there are many of these. Let us start back in the classroom again, as a student might, by asking awkward questions.

PREFACE

On the things every young person wanted to know about physics and was too shy to ask.

Certain axioms of physics, pronounced with great conviction, are very difficult to understand.

For instance – angular momentum is a constant. Yet I can turn around, or put spin on a cricket ball. When did all the turning start and how? Atoms are said to have unitary spin, but who spun them up? And why don't they ever slow down? The earth's rotation is slowing down, but the sun turns faster at the equator than it does at its poles. A neutron star's rotation speeds up when it contracts, but by this reasoning an expanding universe should never have begun to turn at all.

Then there is the statement that light travels in straight lines. This sounds all right until you try drawing a straight line on a piece of paper. You need a ruler to do it properly. You can also walk in a perfectly straight line along a pathway. But how do little particles of light, called photons, manage to travel straight in empty space?

Having formulated universal laws like this, one then finds there are exceptions. Cosmic rays don't travel in straight lines. They just come from all directions, from the sun and other places.

The idea of inertia is embodied in Newton's first law of motion – everything remains in a state of rest, or in uniform motion in a straight line, unless disturbed by a force. Unfortunately no one has ever found anything in a state of rest, nor in uniform motion in a straight line. This is because if you get close enough to objects like this to measure them, you would disturb them.

When you lift something it acquires potential energy, when you push something it acquires momentum and kinetic energy, and this is relative to its place in the universe, its frame of reference. Isaac Newton defined some kind of absolute space, or frame of reference, against which movement could be measured. He performed an experiment by rotating a bucket of water to illustrate this point. At first only the bucket rotated and the surface of the water stayed flat. Then, as the water began to share the motion of the bucket it rose up the sides of the bucket leaving a hollow in the centre. The higher level of water at the sides of the bucket represented an increase of potential energy acquired by rotation relative to absolute space.

If you are in a spaceship in orbit around the earth, when everything is weightless, the same principle applies. An outward, inertial, force will be created when the spaceship rotates, in every respect similar to gravity. It is the same force that keeps motorbike riders glued to the wall of death.

Mach's principle states that these inertial effects are due to the interaction of all the matter in the universe. A rather metaphysical idea. How does all the matter in the universe, billions of light years distant, get to know when your little bucket is rotating? There's more to absolute space than meets the eye!

Isaac Newton realised that if the centrifugal force and angular momentum of stars and planetary bodies was balancing gravity throughout the universe, then everything had to be set turning at the moment of creation, so that the universe would not collapse in one big lump. Newton saw how incredibly difficult such a balancing act would be. After a lot of heart searching he decided that creation must have been arranged by God. He then turned his attention to religious studies.

Einstein's approach was more pragmatic, he simply made angular momentum a constant, so everything had to be turning before there was anything to turn.

When something like a gyroscope turns at high speed its axis gets stuck and when you try to move it, it resists and precesses in another direction. What is your gyroscope stuck in? Don't tell me, empty space is sticky?

A non-rotating mass acts as if all its weight were concentrated at its centre of mass. A rotating mass acts as if its weight were de-focussed along the axis of rotation. The mass can be balanced at any point on that axis, even in the case of the spinning top, when the balance point is outside the mass. Gravity thus seems to be a focussed force which can be partially de-focussed by rotation, and perhaps completely de-focussed by some other means, a form of motion which completely restricted its movement in absolute space, so that gravitation had no affect on it. See Professor Laithwaite's solution in the appendix.

The simplistic solution to overcoming gravity is to project a mass at high velocity. Its path will be straighter if it the projectile is given spin, as with a shell in the rifled barrel of a gun. A projectile travelling faster than 11.2 km/sec, the earth's escape velocity, will fly off into space. At ten times that velocity the mass would proceed unhindered by the sun's gravity and eventually escape from our galaxy in a parabola, which tends toward motion in a straight line. So it seems we have solved the first problem of motion in a straight line and can rewrite Newton's first law as: 'Matter projected at very high velocities tends to move in a straight line uninfluenced by forces'.

Does this new law apply to the motion of light? Not so, because the photons of light, or the little packets of energy which make up the wave, have no rest mass, they only have the energy of their vibrations, which is converted into momentum only when the photon is brought to a halt.

If we could imagine, as scientists once did in the last century, that light was a pure wave motion in a resisting medium, the luminiferous aether, then we could think of waves spreading radially in concentric circles from an origin, the straightness achieved by a linking of force patterns.

There are very few examples of straight lines in nature. The largest ones are those due to linking of crystalline structures. Examples of these are the Giant's Causeway near Portrush in Northern Ireland and at Fingall's Cave off the Island of Mull in Scotland. But is it likely that very large crystals are possible, and that this is the key to the emergence of order out of chaos?

Light rays are refracted when they pass into a denser medium, just as if they were waves. The different frequencies are slowed down by different amounts, the lower frequencies of least energy are bent most and the light separates out into a spectrum of colours. We can therefore see the red colour of the sun bent around the horizon at sunset when it has already set. This is in accordance with the principle of least time.

Two very clever Frenchmen – De Maupertois and

Fermat, formulated the laws of least energy and least time. These laws apply to all reactions in nature so that you can count on them to incur the least action and the least time. The path of light waves is one instance of the laws. Both chemical reactions and celestial mechanics always occur in the most economical way, and you can make predictions about it by Lagrange equations. Furthermore these laws cannot be explained, they are metaphysical.

The laws of least energy and least time also apply to general relativity. This theory offers us another kind of bending of light rays, due to passing near a large mass like the sun, because large masses are supposed to cause a bending of the space around them and light rays thus follow the path of least action through this kind of space. Even solid objects are affected by this bending of space which is postulated to be the cause of gravitational attraction.

How can you bend nothing?

I have here a piece of space – rather like the silly old emperor's suit of clothes. Here is a lump of matter, and hey presto! Your space is bent. This can be proved by radar as well as light waves. Only a fool would deny this evidence. Only a fool would see that the king was naked.

I am that fool who cannot understand how you bend nothing. Space must have a structure in order that you can bend it.

If you don't like the idea of force being caused by the curvature of space you are left with something equally impossible, it's called 'action at a distance'. Isaac Newton didn't like it when he first proposed his theory of gravity. Yet we can see it happening as a magnet will pull something across a table top, with nothing in between. I can see the repulsion and attraction of pieces of paper by electrostatic force, and the way my sellotape twists like a live thing when I try to stick something down.

Two bodies can exhibit an apparent attraction by shooting part of their mass in the opposite direction to the other body. Two bodies can repel each other by exchanging a bit of their mass. But for reasons explained in the book such processes are impossible to observe.

The idea of having lines of force wrapping around an object and pulling it towards you doesn't help, even if you can plot their position around the poles of a magnet with iron filings.

As a boy when I was in my bath I could pull out the plug and make my rubber duck move towards the plughole. Sucking space in to make a force of attraction seems more believable, if there was something to be sucked in.

Fortunately this question is being taken seriously at the time of writing. Experiments are going on at the Max Planck Institute in Germany and at the Ecole Normale Superieure in France to determine whether empty space is really empty or seething with activity, as predicted by the quantum field theory. The experiment consists of moving a very low energy atom, excited by a laser, through a vacuum in a hollow metal cylinder, which is screened by cooling it to near absolute zero. According to the theory, although there is nothing in the cylinder, the atom should bounce out of it. The property of a vacuum, according to the theorists, is that there is always residual vibration, even in a complete vacuum, a background of waves which will respond to anything entering it.

The outcome of the above experiments should prove one way or another my notions about space, that it is not the empty void we once thought, but contains a background of vibrations which cannot be removed and which is the source of all our notions about an eternity outside time.

To summarise the schoolboy questions I have asked:

How did everything start turning?
How does light travel in straight lines in nothing?
How can inertia and angular momentum relate to nothing?
How does the axis of a gyroscope get stuck in nothing?
How do gravitational, electric and magnetic fields attract and repel through nothing?
How do large masses bend nothing?
Nothing must sure be something!

We should perhaps try to understand something about eternity, for we may all be there for a very long time.

PART I

CHAPTER ONE

RE-EXAMINING THE MACROCOSM

In the first part of this book I would like to re-examine large-scale phenomena, and in the second part the microcosmic scale. Certain details overlap but I hope to bring them together in the conclusion.

The problem of determining relative speeds and accelerations depends on the choice of a co-ordinate system. I have already mentioned Newton's bucket experiment in which inertial forces are called into play by a co-ordinate system apparently related to something at rest on the earth's surface. But outside the influence of the earth's gravity a change of course or angular momentum brings inertial forces into play in the same way. In Einstein's scheme for general relativity, gravitational forces were abolished by choosing a co-ordinate system which accelerated relative to the earth, or any gravitating mass, not that it changed the end result. The equivalence of mass with inertial forces was assumed and has been demonstrated by experiment.

Let us look at inertial forces in an illustrative way. Imagine yourself in a jet airliner at 30,000 feet, cruising along at 600 m.p.h. The cloud formations below seem hardly to move and one could imagine oneself at rest, suspended in the atmosphere. One's frame of reference might seem of little importance. Yet it is the thin atmosphere rushing under the wings and fuselage of the aircraft that keeps the plane aloft. A sudden shear wind which removed that lift, in effect changing the frame of reference of the aircraft to the atmosphere, would be immediately apparent to the passengers as turbulence.

Now imagine a flight of aircraft flying abreast, or in echelon. They appear at rest with respect to each other, yet a slight movement of the controls will send one of them veering away, though they remain in line. The initial movement of the controls is all that is necessary to make the one plane recede from the formation indefinitely. That is, if you like, a practical analogue for the action of inertia. This is the idea that one linear boost will cause a mass, unrestrained by forces, to recede from the origin indefinitely. The analogue of the one aircraft receding from a formation is identical if the speed of recession is tiny compared with the airspeed of the whole formation. If, however, it is a larger fraction of airspeed, then other g forces are called into play.

If we now consider the millions of galaxies in the visible universe as a formation of aircraft, what parallels do we find with the above illustration?

For Newton there was only one reference system for the whole universe. For practical astronomical reasons we use co-ordinates fixed to the earth's surface. But if we take a cosmological viewpoint one could use a system based on the ecliptic with the sun at its centre. Going further afield one might use the centre of our galaxy as a reference point. One could also use the cosmic background radiation (an exploding remnant of the radiation from the big bang) as a means of calculating one's velocity relative to space, since as I explain later, this radiation varies in frequency according to our motion relative to it. Or one could use distant galaxies which are receding from us with speed in excess of 33,000 km/sec.* The ultimate fixed point however, from which all measurements could be performed, is the big bang, the point and moment when the universe was created.

All existing theories presume that the big bang completely dispersed into space all the matter and radiation at the scene of the explosion. However, explosions that we see remnants of in our universe seldom follow this pattern. The central object remains after supernova explosion, transformed into a neutron star. Suppose that the central object which produced the big bang is still there, still radiating gamma rays. Such an object would demand, by its presence, to be the centre of our co-ordinate system, even though everything would be receding from that point with very great speed, approaching that of light. Did the universe ever slow down? If not we must begin to look at our reference system in a new light, as if we were on an aeroplane moving at the speed of light away from our origin. The repercussions of such an idea are quite shattering for big things as well as little ones, but I hope to deal with the little things in the second part of this book.

It has been worked out that our galaxy is moving with a velocity of 600 km/sec relative to the cosmic background radiation, some other galaxies appear to have streaming velocities as high as 1,000 km/sec relative to the same coordinates. But are these inertial type speeds like the veering and swerving of a formation of aircraft. In which case their true velocity relative to absolute space could be very much higher. What could cause this veering and swerving? Well, space might be very turbulent, like it would be in the

*The Motion of the Galaxy through Space**

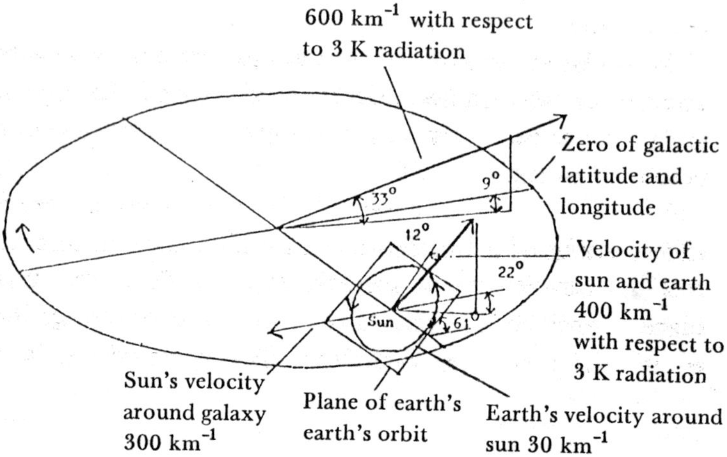

The diagram shows the absolute motion of the earth. The earth travels in its orbit around the sun at 30 km^{-1} and is being swept around the centre of the galaxy at 300 km^{-1}. The earth's speed through the 3K radiation is about 400 km^{-1}. The vector of the earth's net motion lies in the same plane as its orbit around the sun, and is the plane in which the other planets revolve as well. It is an angle sharply tilted upwards (northwards) from the plane of the galaxy. In this diagram the vector of the earth's net motion is depicted as an arrow centred on the sun and pointing upwards and to the right, since the two bodies travel together. Both are being carried by the galaxy's own motion through the 3K radiation. In order to account for the earth's motion with respect to the 3K radiation, the galaxy must be travelling at about 600 km^{-1} in the direction of the arrow centred on the disc of the galaxy.

The diagram shows how vortex streaming around the rim of the galaxy could produce angular momentum over a long period of time.

aftermath of a big explosion, which isn't always a perfectly well-regulated and smooth type of expansion.*

Why would we say that the expansion of the universe might be turbulent? One reason is that far away in space, astronomers have found quasars and galaxies close to one another, that is, at equal distances from us. Quasars are thought to be an earlier stage of the evolution of a galaxy which has an immense output of energy. Yet these objects should both have been created together. Why has one evolved faster than the other? According to special relativity a clock runs slower when it is travelling at great speed. So, if the quasar was travelling at near light speed at some time its evolution would have been slowed down. It now finds itself in the company of an evolved star system which travelled more slowly, by different routes due to turbulence. If both objects travelled with near light speed relative to some point in space only a slight difference in their speeds would produce large changes in the progression of time. The slowing down of clocks in special relativity theory is called 'time dilation'. The difference in speed between a dilation of 3 times and 4 times could be only 2.6% at near light speed.†

Of course, there may have been more than one big bang! As we shall discuss later.

Another curious fact which has been recently discovered is that many galaxies are distributed along spherical surfaces, like the skins of huge bubbles. The saying now goes that space is 'frothy', or 'lumpy'. The insides of these huge bubbles are 50 h-1 megaparsecs across and are not filled with anything we can see. If the matter inside these bubbles was spinning around too fast it may not have condensed into stars, remaining as a dark nebula of gas and dust. But there may not be any matter inside.††

Scientists favour the dark matter idea because one has to explain the gravitational force which is pulling our galaxy along with this huge velocity of 600 km/sec and other galaxies, too, towards places where there is nothing to be seen.

The alternative is the idea of turbulence which I have already outlined by the analogy of the formation of aircraft flying along in company in a clear blue sky. Minute variations of the upper atmosphere can cause them to swerve and veer. Once the initial boost of turbulence occurs the plane will continue to go off course unless this is corrected by the controls. There are no controls for a galaxy! It is borne along like straw in a gale.

The inertial frame for the movement of the galaxy is still related to the origin of the big bang. Its speed relative to that point may be near that of light, 300,000 km/sec! So the veering off course at 600 km/sec is insignificant by comparison and does not require the existence of immense amounts of dark matter.

In the next chapter let us examine what happens when one travels at near light speed, because this shows us what eternity may be like.

CHAPTER TWO

RELATIVE VIEWPOINTS

Many science fiction writers have attempted to describe what the view would be like from the observation porthole of a spaceship travelling through 'hyperspace' at speeds greater than that of light. Most of them imagine blackness outside even more intense than that of the night sky with a sprinkling of stars. One more informed description comes from Fred and Geoffrey Hoyle in their imaginative novel *Into Deepest Space* (Heinemann, 1975. ISBN 0434 34927 5), in which the consequences of special relativity theory were dealt with in a straightforward manner.

As a practical proof of what these writers say in their book there are the observable effects of near light speed in the oft quoted case of muons created by cosmic rays in the upper atmosphere of the earth. The very short lifetime of the muon is 2.2/1,000,000 seconds. At the speed of light it could only travel 660 metres before expiring. Yet these muons actually impact the earth appearing to observers on its surface to have lived for very much longer. However, someone travelling with the muon would not think they had lived any longer – they would instead believe that the earth's atmosphere was foreshortened. These relative effects are due to the consequences of time dilation already mentioned.*

By the same line of reasoning the observer in the spaceship would see the view outside gradually change as it accelerated towards light speed. The colours of objects ahead would move towards the violet end of the spectrum, while those seen astern would shift towards the red end of the spectrum. This a consequence of the Doppler effect, which also explains why the whistle of a train passing through a station falls in pitch when it passes us.

As the spaceship neared the speed of light two black circles would appear ahead and behind, which would grow in size until the whole of space would apparently be compressed into one plane forming a skirt around the ship at an angle of 45° to its path. What would have in effect happened is that one of the three dimensions of our world, the line along which the ship was travelling, would have collapsed, and the occupants of the spaceship would feel themselves to be in a two-dimensional world. In this state they could be instantly transferred from one end of the universe to the other. The snag is, of course, as mentioned in Hoyle's narrative, untold ages would have elapsed on the earth they left behind on their short trip.

The shape of the two-dimensional world of these space travellers would not however, be completely flat. Owing to the transit time of light along all points of the surface to their position, they would be at the apex of a time cone, and their world would cover the surface of this cone, slipping away from them in all directions. The curvature of this cone would enable them to see that there were higher dimensions in their world. They would also notice a stream of events passing across the surface of the cone, seeming to arrive from all directions at once, focussing momentarily, and then vanishing in all directions.

If they attempted to signal to the three-dimensional world they would find great difficulty in doing so. Although light leaves the source with unchanged velocity, no matter how fast the point of emission is travelling, nevertheless a light signal going forwards from the spaceship would have its frequency raised by the Doppler effect. So much energy would be required to send it that it could not travel forward. Likewise, signals received from directly ahead would be shifted into the gamma ray region and produce atomic transformations. Conversely light signals sent backwards from the ship, or received from that direction, would have their frequency reduced to nearly zero, and so become undetectable.

Let us now use the criteria of this spaceship exercise to re-examine our own position in a three-dimensional world. We, too, can only see events along a three-dimensional time cone, and those events pass across it from the future into the past, following an inexplicable arrow of time. Certain phenomena, with very high energies, like cosmic rays, appear to arrive from all directions at once, these are the signals from directly ahead of us, Doppler shifted into the high-frequency end of the energy spectrum. Events from our past are undetectable, being of very low frequency. So, in all directions, we see cosmic background radiation, a remnant of the big bang, which may have been reduced from gamma ray frequencies to radio waves by the billion-fold Doppler shift due to our velocity through space. One might suggest therefore that our world is like a spaceship travelling at near light speed through a fourth dimension of time, so that this fourth dimension has collapsed to a mere surface from our point of view.

There is perhaps the genesis of an extra postulate of

equivalence for both special relativity and the general theory which could be stated as follows:

> A system of N dimensions having a timelike, unidirectional stream of events passing through it, is indistinguishable from a system of N + 1 dimensions travelling at near light speed.

Such a principle would apply as we have seen to the spaceship travelling at light speed and thereby restricted to only two dimensions. It could also apply to our own general situation on a planet, moving around a sun, which is itself in orbit around the galaxy. The galaxy as a whole is part of an expanding universe and moving away from the origin of the big bang with near light speed. In the latter case it is necessary to substitute a term like 'proper time' for the stream of events passing through our world from an invisible fourth dimension, this helps to define the reference frame with respect to the origin of the universe. Movement away from such an origin through a four-dimensional reference frame will thus constrict our world to a thin slice of time as the fourth dimension, which for our purposes has collapsed and ceased to exist. Let us consider the consequences of applying such an idea with the context of special relativity theory. Not least of these considerations is whether time can legitimately be considered as an ordinary dimension.

In Hoyle's book their journey to the end of the universe at near light speed took only a few years, since the whole vast extent of space was compressed by that incredible velocity. If the same criteria is applied to a world of four dimensions then one would have to imagine a cosmos of infinitely greater extent than the one we can now measure. The 20,000 million years which appear to have elapsed since the big bang would have to be multiplied by an almost equal number to arrive at the true duration of the time dimension in our universe.

Something may now tell you that this reasoning is false, that maybe special relativity only operates within a world of three dimensions, that the fourth dimension is something else which exempts one from applying the same rules to it. I shall develop such ideas in the second part of this book, but for the present have to continue to refer to time dimensionally by convention.

If you have gone with me thus far you now begin to understand the idea of an eternity outside time which follows from this reasoning. Things which arrive from outside our three-dimensional time cone appear to come from all directions at once, or to be distributed equally in all directions.

There is also a spacelike dimension outside our time cone which exists forever and ever.

But are there other ways of thinking about time?

CHAPTER THREE

HUBBLE'S LAW

Up to now I have been dealing with time as a dimension like any other, but Hubble's law links time with velocity and energy in a new way.

It was Descartes who devised the geometrical system of measurement we use today – Cartesian co-ordinates. But it was Descartes who also postulated the existence of two kinds of space, the res extensa of extended physical space and the res cogitans, or non-extended mental space. Could there possibly be a res tempis which we could substitute for the second, non-extended form of space which would enable us to deal with the new idea about 'non-locality' which I shall be discussing later.

Ever since special relativity was enunciated, theorists have been concerned with geometrical interpretations of cosmology, and systems of higher dimensions. The equations of general relativity which were meant to account for gravitational attraction and the shape of the cosmos in new ways lead to the conclusion that it might be unstable and in a state of expansion. The astronomer Hubble confirmed the mathematical assumptions of general relativity by discovering a generalised expansion of the universe. This was based on the fact that the spectra of distant galaxies showed a progressive shift towards the red end of the spectrum which was related to luminosity and therefore distance. Hubble's famous law which links recessional velocity to distance, which works out at about 75 km/sec per megaparsec has seemed an extraordinarily exact law to apply to a cosmos which exhibits many chaotic features. But of course, the basic premise of general relativity is that space is homogeneous and isotropic – the same everywhere.*

Evidence is being accumulated which suggests that space is by no means the same everywhere, so laws which are based on uniformity out to the limits of vision could be suspect. One feature of the law which it would not be possible to challenge is that there could be a gradual increase in entropy in relation to time, energy processes do run down, and however gradual this was it would become noticeable over long periods.

It would then be argued that the change we see is exactly the opposite of that one might expect of entropy. The atomic processes we see today would be creating greater energy than they were in the past, everything is speeding up. If processes were running down then the distant galaxies would show spectra shifted towards the violet end of the spectrum.

Some theories have attempted to deal with this, notably A. Milne's called kinematic relativity. These however have not proved popular. The theories were complicated and science favours the simplest solution. But we do now have the idea of a quantum field, which I mentioned at the beginning. This suggests that space is not empty but alive with energy. This quantum field is, if you like, a substratum on which the physical world rests. The creation of waves in the physical world must be carried out against the resistance of the underlying field. The field consists of energy which is subject to entropy, this means that, in time, it will offer less and less resistance to the creation of physical waves. As time goes on therefore the same amount of energy will create waves of higher frequency in a less resistive medium.

Such an interpretation of Hubble's law would of course mean that there was no expansion going on in the universe, but it does not imply that the universe is static, everything in it could be moving more or less in the same direction, at more or less the same speed, as I have already suggested. It does however change our ideas about the shape of the universe, as we shall see later.

Linking time to entropy in this way makes it a material process. We are very accustomed to thinking of one moment of time as identical with any other, all formulas involving time are based on this assumption, that time stands outside any physical system and must be regarded as something comparable with an invisible geometrical coordinate.

But supposing that this was not the case, that instead the arrow of time did involve some irresistible change in the energy substratum. This becomes more credible if we consider the weak force of radioactive decay which occurs naturally at rates so constant that our most accurate clocks are based on these decay rates. The decay of larger atoms can be seen as a consequence of the weakening of the energy substratum of the cosmos and points down the road of the future towards the end of the physical world.

CHAPTER FOUR

COSMOLOGICAL PERSPECTIVES

Looking up at the night sky we can see about 3,000 stars with the naked eye, but the overall darkness of the sky is a curiosity. Around 1800, the astronomer Olbers stated his famous dilemma – that if stars were distributed throughout space to infinity then the whole of the night sky would be filled up with their minute points of radiance, so that the sky would appear as bright as the face of the sun. This dilemma remained with us, in spite of the possibility of there being obscuring dust clouds, until, as already mentioned, Hubble discovered the generalised expansion of the universe, which would take the more distant star systems beyond the point of visibility.

The expansion of the universe reddened the observed light of distant star systems. As the velocity of recession becomes an appreciable fraction of that of light, then in theory, the effects of special relativity become important and the Doppler effect is amplified. At 0.6c for instance the wavelengths of all light emitted from the receding source are doubled.

Tracing the history of expansion back in time is complicated by the relativistic effect, but it leads to an epoch about 20 billion years ago when all the matter in the universe must by this theory have been concentrated at one point in space. At this time it is said there must have been a big bang which blew all this matter apart. Such a violent explosion would have ejected matter in all directions at near light speed. If this was the one and only moment of creation then there should have been nothing in front of the expanding radiation and matter to slow it down, indeed the uniformity of Hubble's law throughout time assumes that matter has maintained a constant velocity from the time of creation.

By the same reasoning, as already postulated, the fireball of the big bang itself would be red-shifted from the gamma ray spectrum to a radiowave a billion times longer, so that from the outset it would appear as we now see the cosmic background radiation.

The point which excites cosmologists at the moment is whether the universe goes on expanding like this forever, or whether the mass of the galaxies is enough to slow the expansion down by gravitation to the point where it stops and goes into reverse, collapsing again into a singularity. For those that hold to the general relativity solution, which abolishes gravity, the path of the expanding mass would be curved in hyperspace until it arrived back at its origin without ever having slowed down.

If the expansion did start slowing down then some of the nearer galaxies would appear to stop moving away from us and even be seen moving towards us, their light would be blue-shifted by the Doppler effect. This has not happened yet – Hubble's law is assumed to be linear throughout time.

Listening to this debate the layman could be forgiven for asking why it is not possible to see in some part of the sky a slightly brighter patch, as a relic of the big bang. The

Doppler Shift at Various Epochs

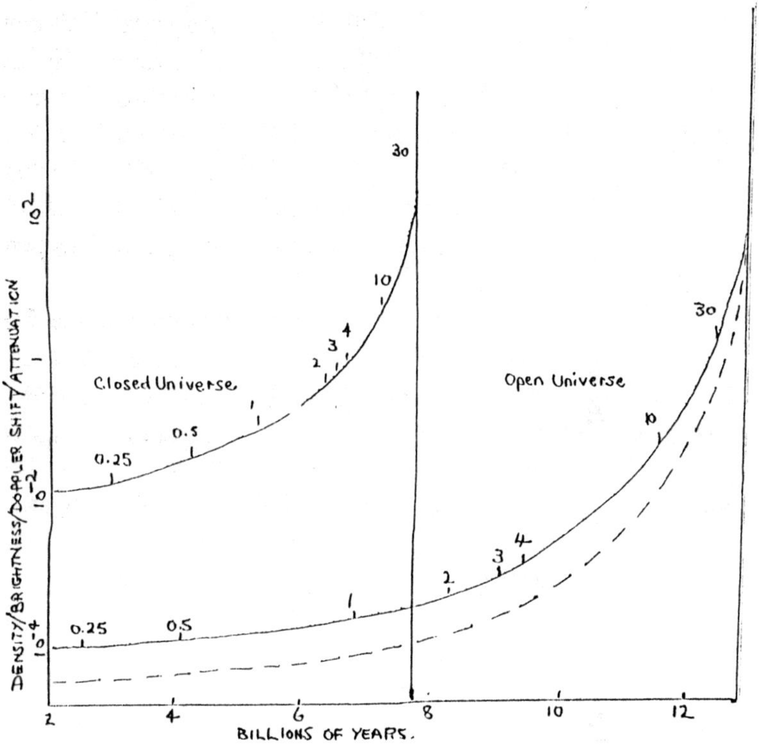

Solid lines show past density/luminosity of universes for open and closed models. For constant CBR (cosmic background radiation), the dotted attenuation line must stay below the density/luminosity at a fixed interval (logarithmically adjusted).*

explanation given in answer to this question is that we live in a spacetime continuum of four dimensions. We can only see events along a three-dimensional time cone through that continuum, and since a fourth dimension has no correlate in our world, past events are seen equally well in each direction in our three-dimensional world. The big bang is therefore spread over the whole sky.

What we actually see is a relic of the big bang called cosmic background radiation or CBR which covers the whole sky and corresponds to a temperature of 2.70K or radiation with a peak at a wavelength of 5 millimetres, having the profile of what is called a Planck curve. There are no spectral lines in this profile of radiation since individual elements did not exist at the time of the big bang.

That is the theory, but more accurate measurements of the CBR have revealed that it is slightly brighter in one direction. This is called the CBR dipole anisotropy. This can be interpreted in two ways. Either it is a Doppler shift, in which case our galaxy is moving towards this brighter area at what is called a 'streaming velocity' of 1,000 ± 300 km/sec, or the brighter area is really the direction where the big bang occurred. This would mean (unthinkably for some) that the cosmos is not a uniform shape, or a perfect hypersphere, as is usually imagined.*

Though the hypersphere is the most favoured shape for the cosmos it is only one special case resulting from the curvature of what is called the Robertson-Walker metric. Other shapes for the cosmos have been suggested. A hyper-ellipsoidal one, for instance, in which heightened spots of brightness in the CBR would be visible in two directions. This concept has been tested experimentally with radio telescopes and found to be unsubstantiated.†

As already mentioned, the single prominence of the CBR is also difficult to explain as a Doppler shift, since it suggests that our galaxy is moving in a certain direction relative to space with a velocity which would exceed that velocity of escape from the local group of galaxies (650 km/sec ± 120 km/sec). Only a vast concentration of dark matter in that part of the sky could produce such an acceleration. I have already mentioned the general postulate of relativity which assumes that space is uniform and homogeneous in all directions. Such a postulate would be violated if immense masses existed in different parts of the sky. The concept of a cosmic hypersphere, as a perfectionist mathematical model for the cosmos, is therefore under threat.*†

An alternative to a closed type of universe would be to imagine a form which is open-ended. This would produce the observable effect of a brighter patch of the CBR as we now see it. Since it is already assumed that we witness past events along a time cone of three dimensions within a four-dimensional space, it would be logical to expand this to a hypercone as a model for the cosmos. This would mean that our existing, visible time cone would lie along a three-dimensional conic section of the hypercone. Another 3D section through the vertex of the hypercone would be a hyperhyperbola, having nearly straight sides. This would indicate the preferred direction in which the CBR would be brightened. In other directions the hyperconic sections would have longer paths to the vertex, wrapping around the hypercone. Although the CBR would be seen in every direction it would not be as bright.††

On reflection it will be remembered that explosions of supernovas which have been witnessed in recent times, like the crab nebula, are far from symmetrical, although the body that gave rise to them was probably a sphere. One can observe filaments shooting out of quasars or exploding proto-galaxies, and prominences on our own sun. The reason is not far to seek. Burnt out stars, white dwarfs and neutron stars, which go supernova, have intense gravitational fields constricting the matter within them. Unless an explosion is triggered at the dead centre of such a star, material will be ejected at the nearest point on the surface, like a pricked balloon. The subsequent collapse of the surface, of the star will further assist the ejection of a filament. It is quite possible therefore, that the big bang itself had this character, that our universe was ejected as a filament and subsequently assumed the hyperconic form.

Existing theories all presume that the big bang completely dispersed into space all the matter at the scene of the explosion. Explosions we see in the universe, like the supernovas mentioned above, do not follow this pattern. The central object, a neutron star or black hole, remains. Suppose therefore that the central object which produced the big bang is still there, still radiating gamma rays. This massive object would, by its presence, demand to be the centre of our co-ordinate system, even though we were moving away from it at high speed.*

Another consequence of this hyperconic idea would be that estimates of the age of the cosmos would vary in different directions. Perhaps this accounts for the present difficulties in fixing a precise value for Hubble's constant (estimates vary between 50 and 100 km/sec).

The mystery concerning angular momentum has already been touched on. In a uniformly expanding, homogeneous and isotropic cosmos the generation of spin cannot be explained in any mechanical framework. However, if one assumed that the initial impulse was turbulent and filamentary, the rotation of this filament in a magnetic field is not an impossibility, it is actually a necessity to prevent a gravitational collapse. A fine balance would have been necessary between gravity and angular momentum, and it is in this balanced state that clumps of matter could have come together to form galaxies. Great difficulty has been found in accounting for the formation of galaxies in a uniformly expanding cosmos.

Initial turbulence is also a useful concept when dealing with the genesis of baryons (heavier particles like protons) in the fireball of the big bang. Very high temperatures are required within an enclosed model to create baryons. The energy/temperatures of a proton, for instance, is about 10 million million degrees K, whereas the highest temperatures calculated only a ten thousandth of a second after the big bang are a million million degrees K. In such a model the lighter particles would have predominated and decayed into electrons and neutrinos.* The turbulent model however, offers the possibility of synchrotron acceleration of charged particles, with supercollider zones, where stable baryons could have been generated at much lower temperatures. One explanation for the high energy output of quasars is that they may be the focus of such supercollisions of very high energy streams of particles, a process which has continued over millions of years. In our area of spacetime we still perhaps see proton primaries of cosmic rays as a remnant of the same supercollision process.

A further consequence arising from the smoothly expanding hyperspheric model of the cosmos is that different parts of such a cosmos could never be aware of each other.

If they originated from a point singularity then the hypersphere would have parts within it which were receding from each other at over twice the speed of light. As ages passed a time cone drawn from our position in space would gradually expand at the speed of light, so that we would become aware of what was happening in an increasing volume of space. When this happens it is found that these areas of newly-seen space correspond in content to our own space. The problem here is – how was the correlation of development stages achieved between systems which had no means of communicating with each other.

In the hyperconic model communication between all parts of the system would be possible from the moment of ejection of the filament from the fireball, if the vertex angle of the cone was less than 60°, that is, if the extreme velocity of recession was less than the speed of light. The whole of the ejected filament may be travelling at near the speed of light away from the site of the big bang, but its components have a very much lower velocity relative to one another, and the hypercone cosmos never slows down, its future is open-ended, it goes on and on and on.

While I have been talking about hyperspheres and hyperellipsoids and hypercones some readers may have been trying to imagine what they look like. I'm sorry I cannot help you, your ideas may be better than mine. It's very difficult to extrapolate geometry into systems of more than three dimensions since the extra ones have to be imaginary. Is this a satisfactory state of affairs I ask myself? Could it be that representing the time dimension as a change of energy would be easier for us?

CHAPTER FIVE

TIME AS A CHANGE OF ENERGY

If we consider time as a non-geometric quantity then the distant past has to be represented as having a quantum radiation field of high intensity, which started with the big bang and has progressively run down through the millenia. The predicted future is a lower energy state. The arrow of time thus follows the gradient of entropy with unerring accuracy.

Our material world can now be seen to embrace a contour line on the energy gradient related to the CBR or 2.7° K. There is a scenario of events prescribed for different points on such an energy gradient.* One such scenario depicts the various events following the big bang as follows:

After 1 second	Temperature is one hundred thousand million degrees K. There is a mixture of neutrons, protons and electrons in thermal radiation.
10 seconds	At one thousand million degrees K. Neutrons and protons stick together to make light nuclei.
20 minutes	At one hundred million degrees K. Nuclear reactions are complete.

300 years	At one million degrees K. No more thermalised radiation.
10,000 years	At one hundred thousand degrees K. Radiation density falls below that of matter.
1 million years	At one thousand degrees K. Electrons combine with nuclei to form atoms.
10 million years	At ten degrees K. Stars and galaxies begin to form.

Initially thermal radiation was about one thousand times as abundant as matter, but as a hyperspheric cosmos expanded, and its radius R increased, the density of radiation relative to matter would fall in proportion to that radius, so that after about ten thousand years the radiation density falls below that of matter.*

When the galaxies were formed the proportion of radiation would have fallen well below that of matter, but at this time the condensation of matter into luminous nebulas would have tended to even out the balance.

Since the lighter elements, like deuterium and helium, are formed in this model at around one hundred million degrees K, it is assumed that the proportions of such elements are determined within the first twenty minutes after the big bang and have not greatly changed since then. This scheme imposes a very tight schedule on the creation and the whole course of events which followed. Thermodynamic considerations yield a 25% proportion of helium in a given mass at the indicated temperature and this is tolerant of considerable variations in the density of the cosmos at this time.

Since the same proportion of this element still exists

in the cosmos today this is taken as evidence for the cosmological origin of helium in the aftermath of the big bang.

In the hyperconic scheme all the above ages can be extended by a factor of at least sixteen, and maybe nearer thirty. This increases the thoroughness of the thermodynamic mixing because the parts of the cosmos do not move apart so rapidly. Likewise the initial temperature could have been higher in a narrow filament, allowing for the formation of baryons.

It has also been suggested that the creation of particles could still occur in supercolliding zones long after the big bang, if the critical temperatures were reached. Helium would then again be produced in the same proportions.

An interesting consequence of the turbulence theory is that radiant masses of gas in nebulas would have been formed into characteristic arcs before winding into the shape of a spiral galaxy. Such radiant arcs have now been discovered.

Another light element of interest in predicting the future course of events in the cosmos is deuterium. Unlike helium it is critically dependent on density in the period following the big bang. Deuterium can be destroyed in the cores of stars but cannot be created there in any quantity, so most of this element now present survives from the initial conditions. The ratio of deuterium to hydrogen is fairly accurately known, about two parts in one hundred thousand by mass. Corrections can be made for the losses of both hydrogen and deuterium which may have occurred during the age of the cosmos. Taking the present density of matter in the cosmos as 10 kg per cubic metre it is possible to calculate backwards in time so that this adjusted ratio agrees with that calculated for the initial conditions in the thermal oven of the big bang. This is a truly impressive feat of mathematics. The only things which could disturb such concrete evidence would be the creation of substantial amounts of matter in quasars, many millions of years after the big bang.

In the original hyperspheric model of the big bang the timescale for the formation of atoms, which are neutral electrically, is about a million years. During this time separate, electrically-charged particles will be in a state of rapid motion. This motion will be of the random, thermal type, so that the resulting magnetic fields created by it would be self-neutralising over a small area.

The situation is not similar in the hyperconic model where there is a generalised motion in one direction. At high temperatures the bare nuclei and free electrons would each create magnetic fields around the axis of motion, being driven in opposite directions by an interaction with this field. It is this combination which would result in supercollider zones.

The hypersphere model also assumes that electrons and protons were present in nearly equal numbers and could combine to form atoms. Neutrons also had to be present in a free state to then form deuterium and heavier nuclei. Neutrons are produced from the interaction of a proton, an electron and an anti-neutrino, which would not be such a frequent occurrence as is required to produce heavier nuclei on any large scale. Likewise neutrons do not survive long in a free state, they have a half-life of fifteen minutes. Alll this poses problems in the matter creation scenario.

When matter is produced from radiation of high energy, by a process called photoproduction, particles and anti-particles appear in equal numbers. Initially, therefore, a mechanism must have existed for the separation of these pairs otherwise they would have just recombined into a flash of radiation, since particles and anti-particles annihilate each other when they collide. The existence of a strong magnetic field as proposed in the hypercone model is a necessary condition for the separation and survival of pairs of particles. Given the existence of a varying magnetic field both particle and anti-particle will orbit each other for a short period before colliding. An electron pair will form a combination call positronium, and two protons can be called protonium (for want of a name). If the particle pairs were moving at light speed through a varying magnetic field of only a few gauss this would be sufficient to stabilise their orbits and keep them apart.

It then becomes possible to suggest an exotic process by which anti-particles are converted into neutrons, since collisions between the positronium and protonium pairs, each electrically neutral would result in the following reaction:

(positron + electron) + (proton + antiproton) = 2 neutrons

This is possible because the neutron is its own anti-particle and two units of anti-spin are provided by the orbital energy of the two sets of particles to form anti-neutrinos.*

The preconditions for the elimination of anti-particles might therefore be:

1. A high temperature.
2. A velocity near light speed.
3. An oscillating magnetic field.

All these conditions can be provided by the hypercone model.

As a last observation about the probable cause for the big bang I would mention that it has been suggested by some that the site of this occurrence must have been a singularity at the heart of a black hole in space. Later in this book I will suggest that such singularities are in a state of rapid oscillation and that this oscillation of trapped energy occurs between two points separated by a vast distance in terms of spacetime.

Such a system is only different in scale but not character from possible singularities at the heart of galaxies, which from time to time exhibit the ejection of filaments. Given that in a state of violent oscillation a slight change of phase may produce instability which causes material to be ejected, the idea that the big bang singularity could have been caused by instability on a larger scale is not so far-fetched. This would be a very rare event due to peculiar resonances. But it is nevertheless something that can be comprehended. You will now see that I have opened a book on bets regarding the possibility of further big bangs.

With a fag end cosmic background radiation of only 2.7° K it would seem that our energy cycle has nearly run out of time and that we are probably due for another cataclysmic event. But maybe it has already happened and the message won't reach us for millions of years yet, if ever. Likewise another hypercone filament wouldn't necessarily be ejected in the same direction as ours, so there is hope.

CHAPTER SIX

COSMIC COMMUNICATION

In an expanding hypersphere every point is moving apart from every other point at an equal rate. This does not disturb the linear transmission of light signals. The expansion that occurs in the time lapse of transmission means that they just have to travel further to reach their goal.

In a hypercone cosmos the effect is very different. Light signals cannot go forwards along the axis of the hypercone or back towards the origin, they can only travel along a three-dimensional surface within the hypercone. Even in this restricted zone they apparently cannot travel in straight lines, since the expansion of the hypercone at relatively slow local speeds displaces them in three dimensions while motion at a much higher translatory speed is occurring along a fourth dimension, the axis of motion of the hypercone.

This high translatory speed of the hypercone means that light signals, which have no rest mass, cannot travel as a projectile would from A to B as shown in the diagram. Light has no inertial mass to carry it along with the expanding hypercone as though it were at rest within it. The signal therefore has to be translated from one point to the next through the agency of an underlying energy field, which does not share in the movement of the hypercone.

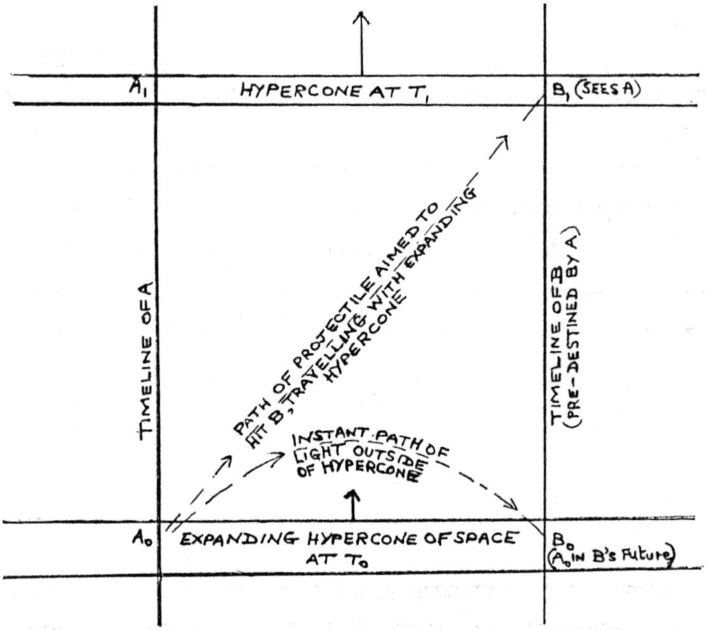

The photon which sets out with the signal passes it immediately to this field which would have to translate it instantly from point to point.

This may sound strange but one has only to consider the way in which wave energy is transmitted in fluids. The molecules of the fluid only move up and down at any point, they do not travel with the wave, they merely pass its energy on to the next molecules. A wave usually spreads its ripples in both directions along every co-ordinate from the origin. In a hypercone however, the ripples cannot spread through the time dimension, only through the three physical dimensions. This is only a restatement of what is already taking place in our world, because we cannot see events happening in the future and past and as yet no theory has been advanced to explain why this is so.

As time passes successive ripples drop out of the physical

cosmos and remain in the energy field that is left behind. Being beyond the influences of physical matter such wave energy is not limited to the translatory velocity appropriate to the hypercone and instant transmission of information can occur between the energy field adjacent to A and that adjacent to B. Were suitable detectors available to sample such a field B could become instantly aware of a signal when it was sent from A without having to wait during the transit time at the speed of light. There is evidence from 'non-locality' experiments, which will be dealt with later, that this kind of instant transmission does occur in a quantum field. (See Appendix Seven.)

CHAPTER SEVEN

THE SEPARATION OF ISLAND UNIVERSES

The diameter of our galaxy is 200,000 light years. It is surrounded by globular clusters which are considered to be its satellites. Our nearest neighbours are the Magellanic clouds, miniature galaxies. Beyond this are members of our local group of galaxies at the following distances:

Andromeda	M31	Distance	2.185 million light years
	NGC221		2.15
	NGC205		2.08
Tri. Neb.	M33		2.38
	NGC147		2.15
	NGC6822		1.533

M = Messier's Catalogue; NGC = Herschell's National Grand Catalogue

The local group therefore appears to be spaced at respectful distances. Galaxies do sometimes collide, but in general preserve their spacing. The almost uniform spiral structure of galaxies suggests some similar characteristic of the force which brought them into being, imposing rotation on the whole mass. Seen from above the galaxy resembles a vortex, rather like water going down a plughole. This has lead to the idea that at the centre of each galaxy is an object like a black hole which sucks matter into it.

The rotation of the galaxy, flinging out vast arms into space is presumably the mechanism which keeps them separate from each other. There is however nothing about gravitation or implicit in general relativity which converts the angular momentum of a body into a repulsive force other than by means of the tides raised on the repelled body.

Einstein originally put into his equations a cosmological constant which gradually increased with distance so that it converts the attraction of gravity into a repulsive force. Although he later cancelled this idea it has recently been revived.

If the energy field is taken into account then angular momentum acting through it could work in the same way as a cosmological constant. A boundary would thus be set up to each star system, limiting the area of space from which it could draw matter. Setting finite limits on gravitational attraction would correspondingly in general relativity set local limits to the curvature of space beyond which it would straighten out again. The result would be an open-ended cosmos, like the one I have proposed, rather than a hypersphere.

The separation of island universes persists through all stages of their development, which one presumes to commence with a high emission quasar, changing gradually to a Seyfert type of galaxy, which have a bright core radiating a hundred times as much energy as a spiral galaxy. Then the usual type of spiral galaxy, followed by the end-phase of an ellipsoid, wherein populations of older stars are found. It is tempting to speculate that galaxies go through a separate life cycle almost in the same way as their star constituents. They would thus not be casual accumulations of gas and stars, but would require an energy system to maintain them, as I shall discuss in the next chapter.

CHAPTER EIGHT

GALACTIC ENERGY SYSTEMS

The greatest problem in any physical theory is accounting for the existence of matter in the form of myriads of particles which exist independently in identical types, with identical properties. Why are some stable and others unstable? Why is the value of their electrical charge, mass, magnetic moment and spin identical? The concept of complete interaction of all particles in the universe has been advanced, but it has not so far been written into the wave equations for the particle.* Reading the theories of physics one gets the impression that each particle can exist in glorious isolation. How would this be possible?

The pragmatist would suggest that an enforcement agency is necessary to ensure the observance of the laws of civilised behaviour in a human community. What sort of agency can one prescribe for elementary particles in space to make them observe common rules of behaviour? Some species of synchronising energy wave perhaps, emanating from a central point and perhaps picked up and repeated by other transmitters at the periphery.

It is easy to speculate that this synchronising energy would have been available in the energy field of the big bang to ensure that all particles created in it were true to their identity. But how has this synchronisation been perpetuated when the remnant energy of the CBR is now so small? There are no free lunches in astrophysics any more than in business. An enormous amount of broadcast energy is required to infuse the whole of space with synchronising waves — such energy as would only be available from the inhalation of large amounts of matter itself.

The source of the broadcast energy must be central to each separate star system so one can only look for it at the centre of a galaxy. Scientific opinion is divided about what can be found at the centre of our galaxy, shrouded as it is by obscuring clouds of dust. The various options are: a complex star cluster, a magnetoid, or an accreting black hole. Whatever it is I would like to suggest that it forms the focus of a broadcast energy system, which supplies a synchronising wave through an energy field not only to the whole galaxy, but its planetary globular clusters, the Magellanic clouds and also links with the emissions of other galaxies in the local group.

The type of object capable of doing this would need to have an oscillating system for converting the raw fuel of matter into an underlying energy field which broadcasts outwards in space to a distance of two million light years. The neutron star, or pulsar, as we already know, radiates sharp pulses of energy. It is suggested that this type of star ultimately collapses into a black hole, as predicted by general relativity. What I am suggesting is that there may be an intermediate stage in this process when matter exists as quarks. This would be the quark plasma star, or quasmar, the idea of which is now exercising the minds of theorists.*

The idea of the existence of quasmars open up a whole new scenario to the progression of events after the big bang. One does not have to postulate the condensation of galaxies from primeval gas clouds. Instead, amongst the ejecta from the initial explosion would have been tiny sparks of quasar-type material, each of which would have been the seed of a quasar initially, and then evolved through accretion of gas clouds into galaxies. So the energising of a fundamental synchronising wave in the quantum field would have been present from the beginning. We could call this a 'matter wave'.

The matter wave would be a modulation of the continuous background of the quantum field which exists everywhere and travelling through it instantaneously. In previous chapters I have argued that the transmission of information through a quantum field can be instantaneous, since it occurs outside the physical space of the hypercone. The regulatory possibilities of the matter wave are therefore not limited by the speed of light.

Galactic Energy Losses

The galaxy is known to be losing energy in several ways – by electromagnetic emissions, by gravity waves and by loss of mass. The loss of mass, amounting to that of 100 suns each year is the most mysterious, there being no explanation for it by any classical theory. The loss of mass is detected by means of a measurement of the increase of the radius of the sun's orbit around the centre of the galaxy.*

This increase of orbital radius could be accounted for by a decrease in the gravitational constant, which is the universal property of matter and produces a force of attraction related to mass. The possible progressive decrease of the gravitational constant with time has been suggested as the explanation for the existence of plate tectonics on the surface of the earth. The decreasing gravitational constant would in time cause the earth to swell, cracking the solid crust and providing the motive force of vulcanism and continental drift. However attractive this idea may seem W.H. McCrea argues strongly against a variation of the gravitational constant, since this would mean not only revising Newtonian mechanics, but there is no formal sense in which general relativity can admit a variable constant of gravitation. Loss of total mass is deemed more feasible as a cosmological effect.*

The mass loss of the galaxy is so large that it implies a continuous process of change within its structure. The total loss over 7.5 billion years would for example, be about 7.84 times its present mass. This is about thirty thousand times as much as the loss through electromagnetic radiation, so matter is not being converted into visible radiation by any kind of atomic reaction such as might occur in stars.

If the lost mass is being converted into some other kind of vibrational field, such as the quantum field, it might escape our attention. Does this enormous outpouring of energy equate to what one would expect to be provided by a broadcast transmitter which was imposing synchronism and identity upon all the particles within the quantum field? If so, what process is involved and what type of object makes use of it?

Gamma rays have recently been detected coming from the centre of our galaxy, which is consistent with the fusion of electrons and protons. This lends credence to the idea that the energy source of the galactic nucleus is a matter-anti-matter reaction producing frequencies which are very high and off the scale so far as our present instrumentation is concerned.

Most physical and chemical processes can proceed in two directions according to the ambient conditions, so the initial reaction which produces matter from radiation can in reverse produce radiation from matter. A very high temperature is the key to largescale reversal of the matter creation reaction – far higher than that which would exist in the heart of even a pulsar, which is the most collapsed state of a star that we know of. Nevertheless we know that pulsars rotate rapidly and have strong magnetic fields. Infalling matter drawn into such fields as we know produces strong excitations of particles, resulting in the northern lights in our own atmosphere and the very high temperatures of the sun's corona, about a million degrees, which is much greater than that of the sun's surface temperature.

Given that the magnetic field of our proposed quasmar was very much stronger than that of a pulsar, due to a high speed of rotation, and that more than one set of poles existed around its periphery, then it is possible that electric arcs would be generated, capable of producing the temperatures necessary for the complete fusion of matter in a species of electric furnace outside the surface of the quasmar.

As infalling matter approached the surface of the quasmar, which would have been reduced to quark constituents, this matter would be crushed down to neutrons. The neutron is its own anti-particle as I have mentioned already, so if huge temperatures existed, as they might in an electric furnace, the weak reaction of neutrons might equally produce: proton, electron and anti-neutrino or anti-proton, positron and neutrino. This you will remember is a reverse of the original creation scenario whereby anti-particles could be eliminated from the process. So matter passes out of existence by the mutual annihilation of particles and anti-particles, and the immense energy field created by this provides a synchronising radiation capable of maintaining in existence all the particles in the galaxy and its environs.

There is a paradox in star evolution that the final crushing of matter into its quark constiuents would serve to prevent random thermal motion. It somehow represents a cooling or crystallisation of the mass which manages to release energy. The paradox is that this penultimate state of matter does not follow the rules of entropy, it becomes more ordered, the whole volume of the quasmar becomes organised rather like an enzyme or a biological conversion mechanism. While we can speculate upon the form the organisation would take, as I do later, it is hard to adduce any proof.

Nevertheless one can speculate upon a new principle relating to gravitational attraction, in that it may not be just a crude collecting agency for gas clouds, turning them into stars, but it also has an entelechy or ordering principle, something that the philosopher Hegel would have liked to ascribe to the absolute.

If this is so, then apart from the existence of rotating galactic systems, and planetary systems within them circling the constituent stars, we should also be able to observe the incipient stages of an ordering process beginning in large astronomical bodies, something related to convection systems along lines of magnetic force exhibited at the surface of the body.

In the case of the earth the edges of tectonic plates do tend to run north and south, following lines of longitude and the earth's magnetic field, so that it might be assumed that the convective system of constructive and destructive margins was to some extent influenced by magnetism.

In the case of the sun, which rotates at different speeds at the equator and the poles, an overall magnetic field would tend to be wound up in time and form vortices. These take the form of sunspots with their eleven year cycle. Investigations into the magnetic field of sunspots started as long ago as 1908 when Hale noted that spectral emission lines from the sunspots could be resolved into finely spaced components, each of which was polarised. The splitting of the lines was called the Zeeman effect, and is due to the emission having occurred in a strong magnetic field. The field strength can be as high as 4,000 gauss, thousands of times stronger than the earth's magnetic field, and may extend over an area equal to the surface of the earth. Such powerful fields cannot be due to magnetic substances embedded in the sun, but must be caused by electric currents circulating in the interior. Within the umbra of a sunspot the field lines emerge more or less vertically, indicating that the convected current comes from deep within the sun's interior.

Just as in the case of the earth, where there are opposed systems of upwelling and downthrusting margins, so also sunspots frequently appear in pairs, aligned parallel to the sun's equator. They are therefore called bi-polar spots. Magnetic measurements reveal that they have opposite polarities, field lines emerging from one spot curve downwards into the other, forming a closed loop. The polarity of the leading spot of a pair is always the same and opposite in the northern and southern hemispheres throughout a solar cycle. Then, at the commencement of the next cycle the polarities reverse in both hemispheres. This has a curious parallel with the magnetic field reversals known to occur on earth at intervals of about a million years.

The magnetic field of a neutron star rotating with a period of millisecs. It emits pulses of this frequency of rotation in the X ray, optical and radio bands, by a synchrotron process from the decaying energy of high velocity electrons.

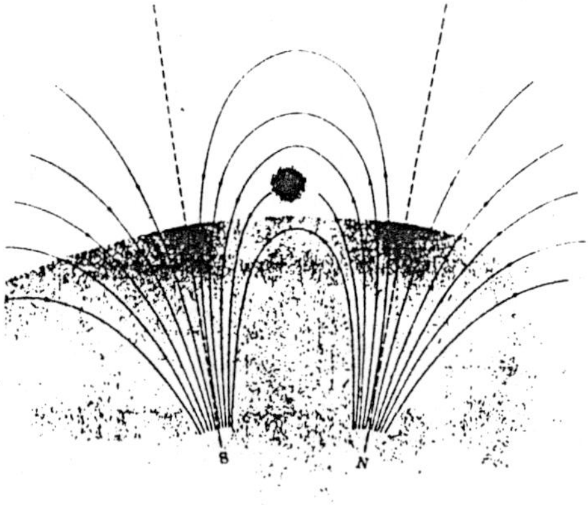

The formation of a loop prominence on the sun from magnetic fields going vertically from & to its surface.

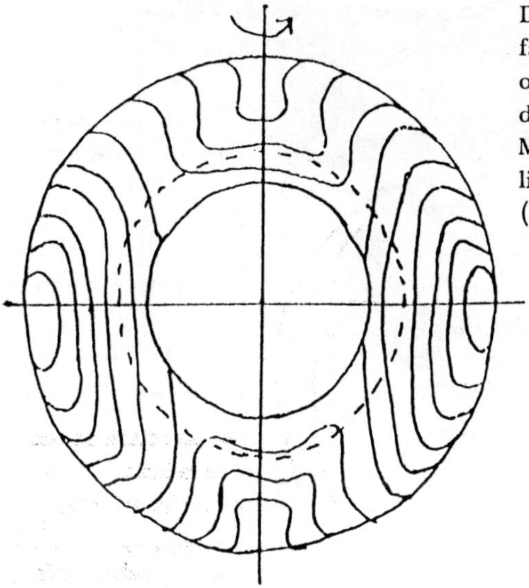

Differential rotation surfaces of the sun in the outer radiative zone down to dashed line. Magnetic activity closely linked with rotation. (Rosner & Weiss).*

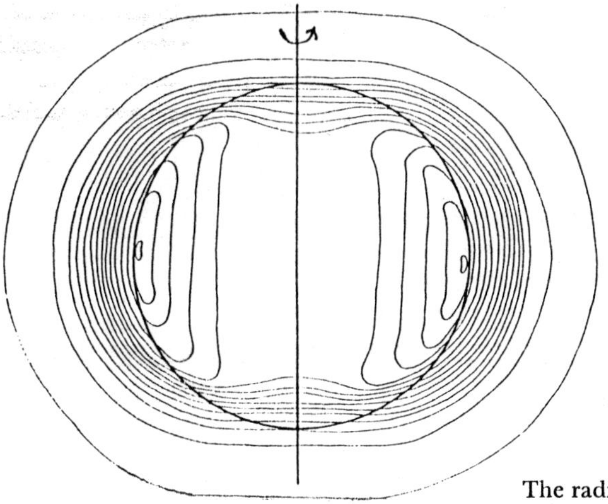

The radio picture of the sun illustrates the linking of magnetism and rotation by the direction of limb brightening.

A computer assisted image of the red giant star Betelgeuse obtained by speckle interferometry. This shows a number of twin features on the surface of the star which could be similar to sunspots and have a magnetic polarity. (Kitt Peak National Observatory, USA.)

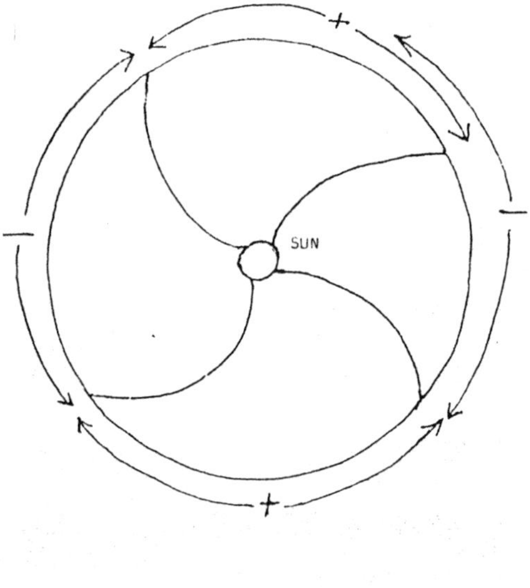

The structure of the solar wind as determined by an interplanetary spacecraft in a 30 day orbit. This reveals heightened charge in the centre of each sector which could predispose the formation of planetary nuclei.

The Magnetic Field of Sunspots

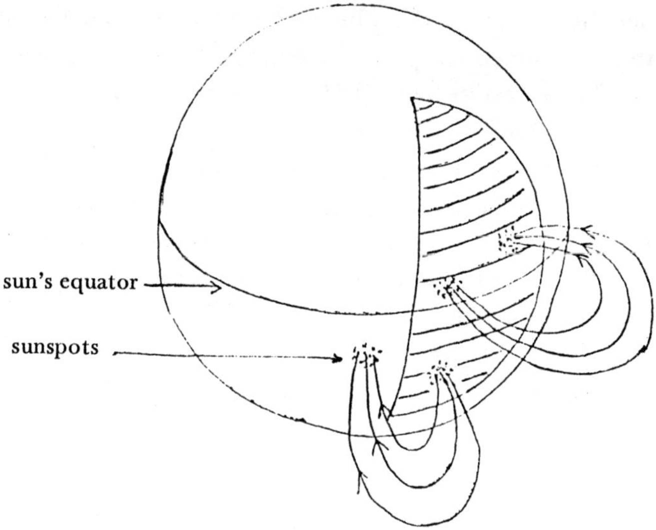

View of Magnetic Field of a Quasmar

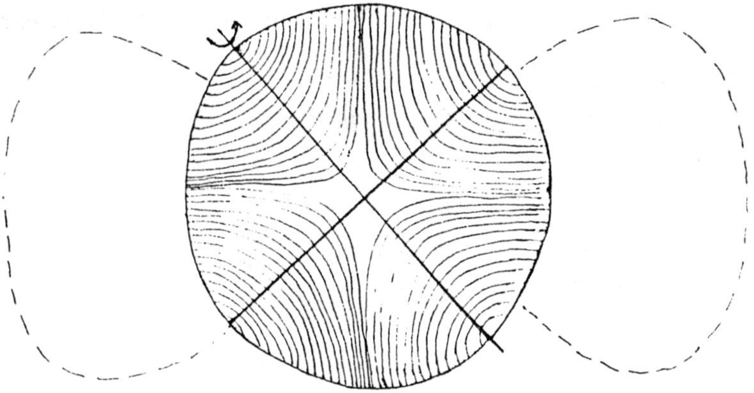

Lines indicate vertical plane of quark plasma. The central region can be considered a singularity in oscillation between this and other regions of spacetime. This would be a logical evolution of the sun's differential rotation zones and magnetic field.

The fact that these kind of magnetic phenomena are not peculiar to the sun is illustrated by a recent photograph of the red supergiant star Betelgeuse, taken with the aid of speckle interferometry. This clearly shows pairs of spots above and below the equator.

The red spot of Jupiter cannot be placed in the same category since it is a single feature. The surface of Jupiter turns faster at the equator than at the poles, like that of the sun. The best estimate of its correct rate of rotation is made by matching the rate of rotation of its magnetosphere, which is slightly lower than that of equatorial rotation.

Jupiter's spot changes in brightness through the years. Material appears to come up in the centre and spread to the periphery in an anti-clockwise rotation. Other temporary spots have been observed on the surface of Jupiter from time to time, notably in 1967, where one lasted about twelve days. The great spot is slightly mobile on the surface, so if it was a kind of volcano it would be a very frightening apparition.

These observations may seem slim evidence to base a theory on, but they are actual observations not just mathematical models. Anyway, let us try to visualise what the structure at the centre of the galaxy may look like.

CHAPTER NINE

THE QUASMAR

The idea that there are massive rotators in galactic nuclei has already been proposed by L. Woltjer.* The magnetic fields produced by such objects are seen to be the efficient source of relativistic particles and to cause the emission of radio waves from vast areas of space surrounding a galaxy. In comparison only 10% efficiency of acceleration can be achieved by a Fermi process of neutral line acceleration.

Woltjer's model was intended to explain the existence of Seyfert galaxies and quasi-stellar objects, where the magnetic field was expected to be complex and unsteady. The order of size of the rotator is given as a billion solar masses, with a rotation period of 12 days and a radius of 3 billion kms. The rotator is assumed to be centrifugally supported against gravitational collapse, nevertheless the possibility of a black hole existing at the galactic centre is not ruled out. This means that the object evolves towards a collapsed state which a further increase of rate of spin and a much reduced mass.

The lower limit at which gravitational collapse occurs is known as Chadrasekar's limit, which is 1.5 solar masses. Stars heavier than this will collapse into white dwarfs at the end of their life. They eventually degenerate into neutron stars which are distinguished by the emission of X rays and pulses of radiation at other frequencies. The gradual process of collapse and squeezing of the constituent atoms by gravity thus results in higher frequencies of electromagnetic radiation being produced. The practical limit which might be reached therefore would equate to the Compton wavelength of particles themselves. The galactic centre is in fact emitting gamma rays consistent with a process involving the fusion of electrons and positrons.

Professor Hawkinge has already pointed out that intense gravitational fields in the vicinity of a black hole would lead to the creation of particles. Due to the presence of magnetic fields an anti-particle might be drawn into the hole and the particle escape. Such a manifestation of particles drains energy from the black hole, leading to its collapse. Overall however, an increase of mass would occur such as to counterbalance the matter accreted into the hole.

In our galaxy one must account for a considerable loss of mass which is not converted into observable visible radiation, gamma rays or X rays. Very high energy gamma rays would tend to produce pairs of particles by interaction with intergalactic dust, which might or might not separate in the magnetic field surrounding the galactic centre.

The possible explanation which I have introduced to explain the mass loss is that it is used to create a stabilising quantum field, a form of vibration akin to sound waves. Such waves would have to be produced by a mechanical vibration of the mass as a whole, random thermal vibration would not do.

To explain this one sees how electromagnetic waves are produced by the movement, collision or annihilation of charged particles. In a neutron star the closeness of the particles effectively neutralises electric charge. The increasing intensity of the gravitational field will reduce the frequency of the radiation which is emitted, and of course, in the case of a black hole, no radiation can escape at all. However, the harder the particles are locked together the more they are likely to vibrate, like a huge crystal.

In addition to the mechanical property of rigidity the object would need to have magnetic permeability, the ability to pass electric current, in order to account for the radio emissions and synchrotron radiation which one finds surrounding galaxies.

The overall charge of a large mass is usually taken to be zero. With the increase of gravitational potential however, a point is reached where charged particles are expelled. Any residual charge will therefore accumulate on the surface of the body, since it behaves like a Faraday cage, in that no net charge can exist in its interior. Small charged particles will actually be expelled by radiation pressure from the surface of a star – this happens in the case of the sun, where streams of electrons are pushed out into space along four distinct ecliptic paths in a cross-shaped inhomogeneity. A magnetic field of this shape therefore appears to exist around the sun, just as around the galactic centre there is a similar inhomogeneity which produces the spiral arms of the typical galaxy.

Radiation pressure will not be as great in a degenerate, or collapsed star, so that electrons will not be ejected into space, and they will be induced to rotate in the cross-shaped inhomogeneity around the object, acquiring synchrotron energy as they do so.

Let us consider what may happen in the core of a collapsed star when the neutrons are forced very close together. When spacings between particles are reduced to the repulsive core of the neutron – about a 1.5 thousand million millionths of a metre – nuclear force will predominate over electric charge. It is at such interfaces or changeover points that oscillation is possible. To take the case of the quartz crystal, oscillation can be produced by an electric potential applied to two particular faces of the crystal. Pressure on these same two faces will also reciprocally produce an electric potential due to the piezoelectric effect. The interface here is that between the ionic bond and the molecular adhesion of the crystal.

Similarly, if very high gravitational force causes neutrons to be squashed into a configuration of their constituent quarks then this would have the same characteristics as a crystal structure. The reason is because quarks have electric charge and could assume an overall, coherent pattern of alignments throughout the mass giving it great rigidity.

It would therefore not be unreasonable to assume that mechanical vibration of the whole mass would occur if electric potentials existed on the surface of the object, providing that they differed from point to point. The process I have in mind is that of sheets of quarks being formed, each of opposed polarity to its neighbour. Now if such sheets ran circumferentially around the object, like the skins of an onion, electric charge would have to be trapped inside the pseudo-Faraday cage, which one knows to be impossible, also the convection of thermal and radiant energy would be difficult, leading to an explosion.

It is therefore necessary for the opposed sheets of quarks to extend downwards from the surface radially into the interior. The constitution of a quasmar would therefore have to be convoluted. Successive radially oriented sheets of plasma, roughly spherical in shape and of diminishing size towards the poles, would have to be axially disposed in three, four or more segments, rather like the armature of a printed circuit motor. Magnetic fields generated by the electric current upwelling or downward flowing at the principle junctions of the sheets would provide the motive power of a vast electric arc furnace above the surface. This would fuel the matter-anti-matter annihilation process between adjacent current vortices, burning off matter with unbelievable fury and converting the energy into the vibration of the whole of the quasmar, which is then radiated throughout the environs of the galaxy as a quantum field, synchronising all the particles therein. We glimpse a pale imitation of these processes in the magnetic storms on the sun's surface, which produce temperatures of one million degrees in the sun's corona.

It is the thickness of each sheet of interlocked quarks which will determine the frequency of vibration of the quasmar. This thickness can only be a tiny fraction of the diameter of the repulsive core of the neutron. This is because one must be at the boundary where nuclear force and electric force are equivalent in order to achieve oscillation, and this is about equal to the Compton wavelength of the neutron.

There are at least three quarks of different varieties in a neutron, so one can assume that it is the intermixing of their respective vibration rates which synchronises the production of various varieties of baryons, hadrons and leptons by addition and subtraction of energies.

The effect of rotation of the quasmar would be to enhance the magnetic force of the annihilation process. Its surface would be crossed longitudinally by innumerable edges of quark plasma sheets, running radially into the core of the object, rigidly fixed in place by the colossal gravitational field. A quantum field vibration sprays outwards from this surface along the plane of the galaxy. This field has that quality of instant transmission to the boundaries of the universe characteristic of phase waves, carrying information rather than substance, in order to achieve the necessary synchronisation of particles.

The normal field radiation emanating from a quasmar would contain a Doppler variation due to emission from each separate limb of the object. The splitting of spectral lines of gamma rays coming from the galactic centre would lend proof of the size of the object. Woltjer's model, mentioned earlier, had a surface velocity of 170,000 km/sec, astonishingly over half the speed of light.

CHAPTER TEN

GALACTIC EVOLUTION

Our galaxy, the Milky Way, is one of an estimated 100 billion galaxies in the universe. It belongs to a sub-class of 30 billion spiral galaxies. A survey of active galaxies and quasars reveals an extraordinary variety of sub-species. In each group there is a type, frequently named after its discoverer, or alternatively the technique of discovery, though some names merely relate to the place where one finds them. The various types are:

Markarian (discoverer): Strong ultraviolet continuum, of two types – spiral and diffuse. Spirals have a bright nucleus and include Seyfert galaxies with broadened spectral lines. They were probably formed in the last 100 million years.

Haro (discoverer): A sub-class of the diffuse Markarian type with intergalactic regions of ionised hydrogen.

Compact (discovered by Zwicky): Having a high surface brightness, barely distinguishable from stars. There are red and blue types.

N – (types): A bright star-like nucleus surrounded by nebulous envelopes. The nucleus has the colours of a quasar and an envelope like that of a normal galaxy.

Seyfert (discoverer): Have a starlike nuclei, rich emission of broad line spectra. Substantial non-thermal emission at ultraviolet. Have colours like a quasar. They are variable over a few months. Forbidden hydrogen lines are produced, violating the Pauli exclusion principle.

Radio type: Are very distant and have a large source of synchrotron emission friom high-speed electrons. Radio clouds suggest a largescale, tangled magnetic field.

Quasi-stellar type: Seen only at large red-shifts. Times of variability vary from a few days to decades. Large output of spectral lines. They have a diameter of about one light year.

The recurrent theme of the above types is that of an active nucleus, and it has become apparent over the years that the central regions of galaxies are the seat of very violent activity. Some appear to be expanding very fast, and even exploding. The energy released by these types is from tens to hundreds of times that of a normal galaxy, and far above that available from normal atomic reactions.

The original creative act of the big bang is supposed to have produced the original elements hydrogen and helium from which our galaxy was born in the proportions of 75% to 25%. At that time, 12 billion years ago no heavier elements were thought to have existed, and no stars.

Much is still unknown about galactic evolution, why for instance, galaxies are not much smaller or much larger than they are. It is assumed that the limits of our galaxy, at a radius of 100,000 light years, including the satellite magellanic clouds and globular clusters, defines a gaseous sphere out of which the various stars are condensed, and are still condensing. Gravitational collapse is supposed to have produced the denser, inner clouds of stars which rotation formed into a disc. After several revolutions spiral arms form in the disc. This took one billion years. Clusters of stars then formed in the spiral arms and our own solar system took shape perhaps 7.5 billion years ago. This must have been after many heavier stars had gone through their shorter lifetimes and exploded as supernovae to scatter the heavier elements which we find on earth.

The Milky Way contains at least 200 billion stars. Our sun is one of them and lies 30,000 light years from the galactic centre, around which it rotates in 230 million years.

The core of our galaxy, forever hidden to optical telescopes by dust clouds is about 13,000 light years in diameter. Stars near the periphery of the core rotate about it every 50 million years. At the very centre, in a region of about 3 cubic light years, the rate of rotation is once every 10,000 years. The most mysterious object in this region is a small radio source with a radius of 1.5 billion km located at Sagittarius A West. At the centre of this radio source there may be a super massive black hole with a diameter of less than 15 million km. But with a mass of 5 million suns. It is the accretion disc around this object which is said to capture and transform the equivalent mass of 100 suns a year.

If my preceding arguments are correct however, then the central object may not be a black hole at all, but instead a quasmar, which devours matter in a different way, not by crushing it out of existence, but by converting it directly into the subtle energy of a quantum field by the equivalent of an electric arc furnace. It uses the enormous electric and magnetic forces of the vibrating quark plasma star to sustain an invisible, yet all-powerful field, which radiates far out into space beyond the remotest satellites of our galaxy. If this is so then the mystery of why all galaxies fall within a certain range of size is solved, for there are obvious limits of size, mass and spin within which a quasmar can function and therefore limits to the distance over which it can transmit.

The immediate mystery is how did such an object come to be formed? Following that of course: how could any sort of matter have existed before it was formed?

In a sense the quasmar must be considered as an end product of the evolution of matter through all its forms. The end must then resemble the beginning, since before the big bang matter must also have existed in a highly organised and concentrated state. What many scientists believe now is that the universe did not expand uniformly and evenly in all directions after the big bang, but that strings compressed by magnetic forces flailed around, more or less at random, producing discontinuities which later condensed into galaxies. If instead of strings one can concede that the big bang threw out sparks of quasmar size, such that it must become the seed of a galaxy, then the picture is complete.*

The reader will have no difficulty in identifying such seeded galaxies as quasars, no more than a light year across. They are distant and their energy source unexplained. But they are observable and not theoretical, like strings. One would then be able to account for the broadened spectral lines characteristic of their emitted radiation, which suggests that rude and incoherent matter in the vicinity was being brought into due form by the synchronising action of their quantum field vibration.

The next stage in galactic evolution would then be the Seyfert type of galaxy, in which high energy output continued, with the same, broad spectral lines, but gradually huge gas clouds and new stars commenced to form in the conventional mould of a galaxy, within the lobes of the radiated field, stretching furthest from the equatorial plane of the quasmar and following the magnetic field inhomogeneity.

The radio galaxy variant would arise where two quasmars in close proximity each produced high frequency energy fields which beat together to produce a different frequency in the radio spectrum.

The evolution of planetary systems, which orbit their primary star in the one equatorial plane of the ecliptic, can thus be seen as the microcosm of the galactic system and suggests that stars act as repeater stations for the quantum field. There is no other explanation either within Newtonian nor general relativity physics for such uniform alignment of orbits.

The next stage of galactic evolution after the spiral form, or barred spiral, would appear to be the oval type, which consists of older stars. The gas clouds have all condensed into stars and the spiral arms have smoothed themselves out by many orbital rotations. A total of four distinct phases can be observed in the development of galaxies, with minor variations in each phase. The question which now arises is whether such phases are linked in a positive way on a single evolutionary path like the Hertzsprung-Russell diagram for the evolution of stars.

We can see the past states of galaxies down the time cone right back to the initial stage of quasar, and there is an apparent conformity of the states in matter along the whole of this timescale. This argues the fact that in addition to a lateral synchronisation of matter through a quantum field in present time *there must also be a similar synchronisation extending backwards, and perhaps forwards in time through the agency of the same field.*

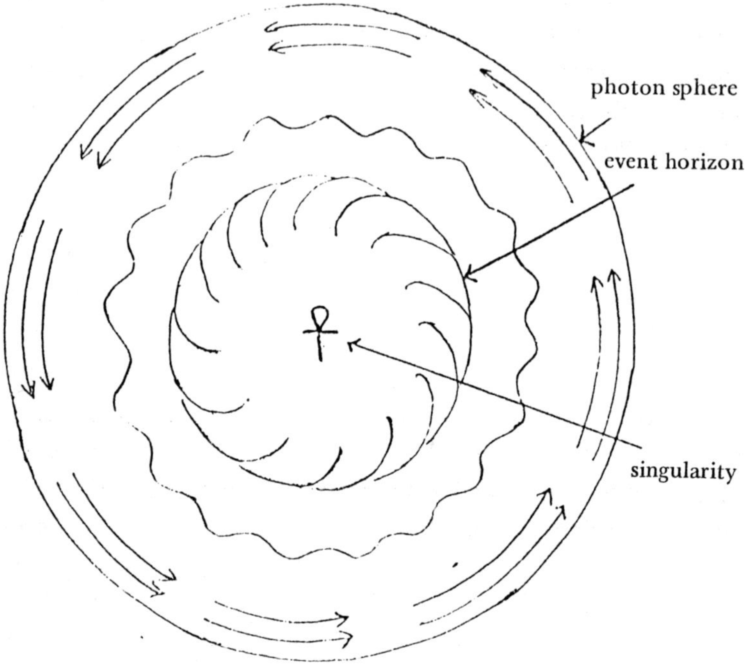

A section through a black hole showing the photon sphere, the event horizon and the singularity.

If one perceives the quantum field vibration as raising and lowering the average median level of such a field, then by the arguments I have so far advanced, and which will be elaborated in the second part of this book, this will alter the time spread of the oscillation. A formation like a quasmar at the galactic centre would produce such an amplitude of oscillation that its effects are likely to be perceived both in the past and the future at distant points where the peaks and troughs of the vibrational energy happened to match the local median level.

For every galaxy there would then exist two correspondence points in the future and the past where synchronising energy appeared. In this theory then, galaxies at various stages of evolution would be chain-linked through space-time, so that they reinforced one another. The positions of galaxies would therefore be related to this synchronising energy, groups of galaxies would be replicated in an earlier cyclic zone and there would be vast regions of empty space where no synchronising energy appeared. This is of course, the situation we find and so the theory responds to observable conditions. Moreover, the spiralling of galactic formations produced in the hypercone type of universe would introduce communication between galaxies above and below one another on adjacent turns of the spiral.

Lest this seems too much of a theoretic leap we can perhaps relate it to existing general relativity theory. According to this theory there is no limit to the amount that space can be bent or warped by the presence of mass. I do not take issue with the bending, but contest the idea that this could squeeze atoms together until they disappear. This requires FORCE and as I have made clear before, there are no free lunches in astrophysics. If force has to appear then something must provide it, one cannot tap an infinite source by mere mathematics.

However, disregarding my doubts, by general relativity theory the process of collapse goes on in burnt out stars above Chandresakar's limit until they finally become a black hole in which matter is crushed out of existence. The German astronomer Schwarzchild discovered the mathematics of this idea after the general theory was published in 1916. Schwarzchild had to assume that the collapsed star was perfectly spherical and non-rotating. Near the star he found a number of strange conditions which did not apply in Newtonian physics.

General relativity prescribes that time runs more slowly in a high gravitational field, so moving still closer to the black hole one next reaches the event horizon. This is where time stops and one can no longer communicate with the outside world, since any such communication would take an infinite amount of time. The traveller has in all respects disappeared from our universe.

Finally, at the centre of the hole one reaches the singularity. Here exists a tiny point where space is so bent that anything which approaches it is crushed out of existence. It is like a whirlpool in space, devouring the matter that spirals in towards it, drawn inwards by the steepening sides of the gravity well around it.

But what happens to all this matter that is crushed to nothing? It cannot escape outwards again, even as radiation, since the steep sides of the gravity well would reduce the frequency of the radiation to zero. What happens also when the sides of the gravity well become vertical, warped at 90°, so that they simply cannot generate a higher potential force.

Warped Spacetime

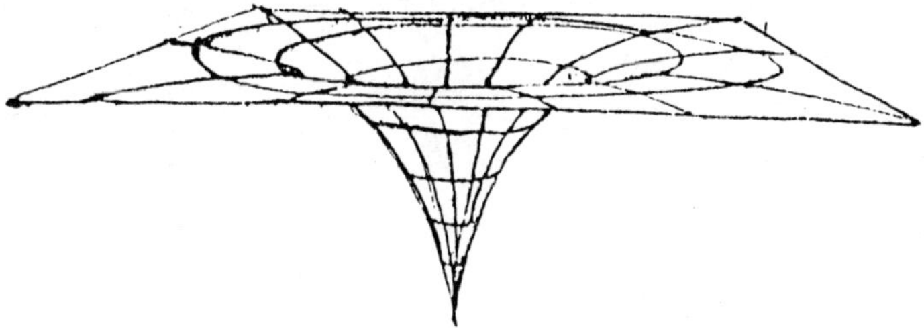

The four-dimensional space near a collapsed star is distorted and the slope of the well indicates the high accelerations of matter falling into it.

The Einstein-Rosen Bridge

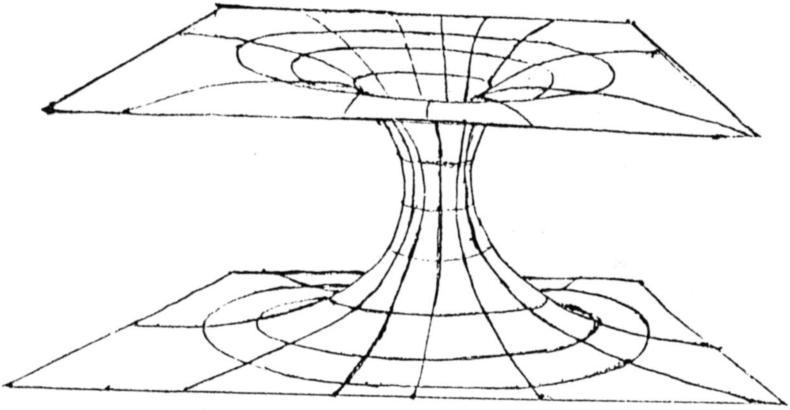

This diagram shows the geometry of spacetime about a non-rotating black hole. The hole through the centre of the well connects two separate universes. The event horizon is at the centre of the hole.

The Einstein-Rosen Connection

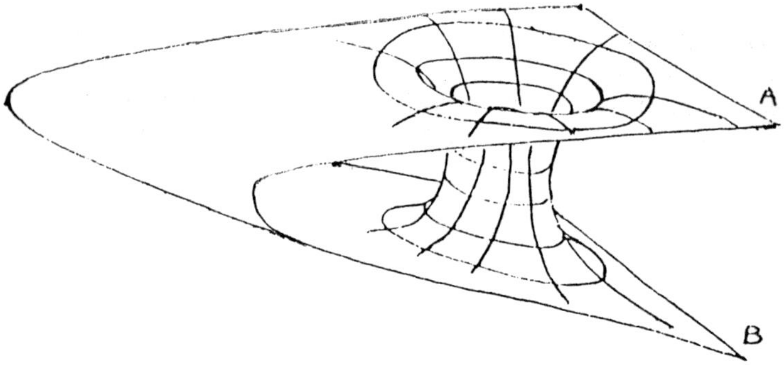

Two distant points of the same universe connected by a shorter route through the hole.

The Wormhole

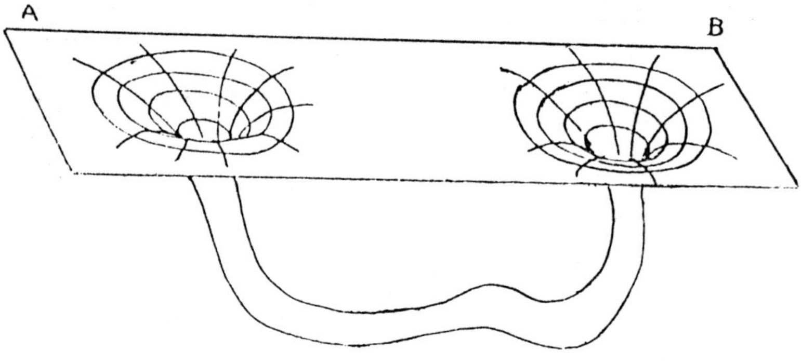

An unfolded version of the previous diagram. The distance along the flat surface would actually take much longer than a trip through the hole where time stands still.

The solution found by Einstein and Rosen in the 1930s was that, after passing through the point where the sides of the gravity well became vertical, they opened up in the opposite direction, so forming an inverted dish, joined to the underside of the first one, which was the original black hole. The point, or hole, where the two dishes joined, was called the Einstein-Rosen bridge, or more picturesquely, a wormhole. Any material thing, or energy, passing through this wormhole, would emerge into another kind of universe in which everything was upside down, or back to front, or of the opposite polarity of electric charge.

What is now generally theorised is that either these bridges do connect with new universes or that they connect with distant parts of our own universe through a region outside time. If the points of entry and emergence were opened out on a plane, then it would seem they were connected by a kind of pipe, or wormhole, outside space, through which travel was instantaneous. Matter would emerge with great speed at the other, now called a white hole. This means that there would have to be one-way traffic through the bridge, otherwise collisions would occur and the bridge would be destroyed.

The reader will have begun to discern the possible outline for a creative plan for recycling garbage from an old galaxy through the wormhole and using it to reform the genesis of a new universe around the white hole. White holes could further be identified with quasars at some remote part of the universe. This would be a continuous process of destruction and recreation, exactly in balance and part of a harmonious total scheme of things. It is much simpler to account for, and more efficient, than a big bang.

But before we get carried away with such an idea let us

stop for a moment to consider how the energy passing through an Einstein-Rosen bridge could possibly escape on the other side. True it would rush into the black hole with great speed, but when it tried to emerge from the white hole it would be restrained with equal force, pulled up short and fed back into the singularity. Theorists have suggested ways of building up synchrotron energy before plunging into the black hole, sufficient to escape on the other side, but these are very special paths of lucky coincidence, not likely to be followed by the majority of material. The majority would thus be trapped in the bridge and finish up oscillating back and forth through the wormhole never getting out.

Clearly the oscillatory solution is the one which has to be considered as the logical outcome for the massive nucleus of a galaxy, and the resulting energy of vibration of the quantum field is the means by which this accreted power is dissipated from the nucleus.

Nevertheless it can be seen that some phenomena related to the quasmar could occur at a distant point of the past and the future, where the turns of the time spiral cross above and below in terms of the median energy of the quantum field.

The scheme now proposed is the twinning of a galaxy with a quasar at the higher median energy point and with some as yet unknown object on the turn of the lower median energy point.

On the opposite side of the evolutionary cycle the older, oval type galaxies would be twinned with Sefert types, and all phases of the evolutionary cycle would be in between on the the spiral.

The scheme I have suggested will be seen to be radically

and their linking through the activity of a convular nucleus, which extends through the bilateral quantum field.

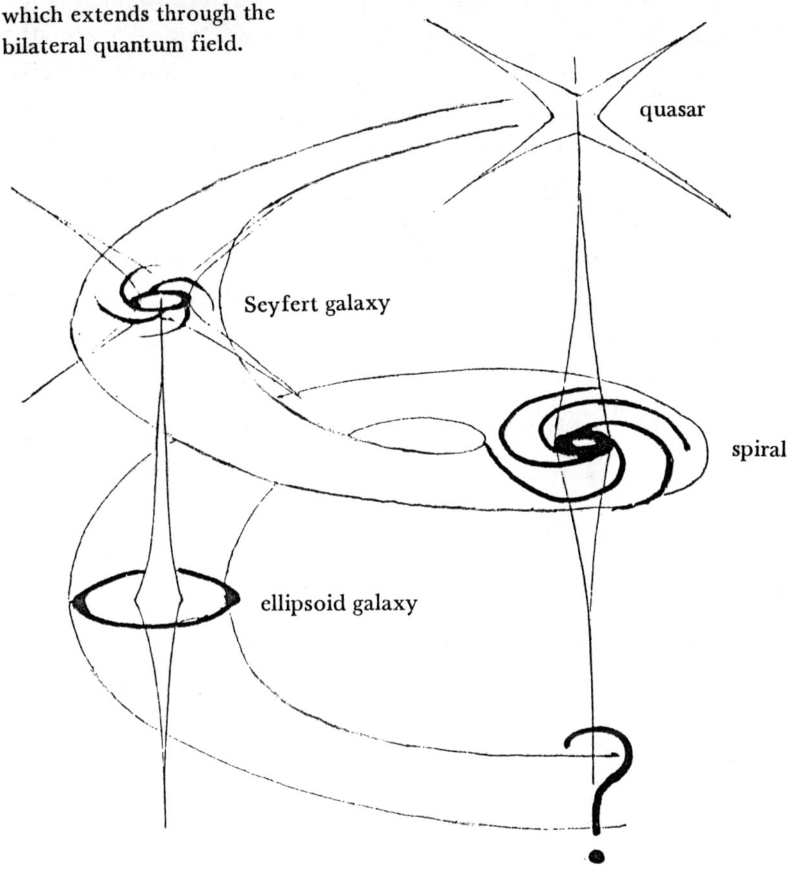

different from present ideas about open, closed or cyclic universes. The rate of expansion of a hypercone is slower than that of a hypersphere, so gravitational condensation is more certain but equally balanced by rotation in a spiral fashion. Matter collects into galactic groups from the beginning around the initial sparks of quasmar material ejected from the big bang. These quasmar nuclei transform and synchronise gas clouds around them. As they evolve they reach through time in both directions to seed and synchronise new quasars and link with older galaxies. The whole process is not cyclic, since it requires only the initial impulse to set it running and from then on it is regenerative, like a living organism.

Some thought therefore needs to be given to the teleology of the hypercone.

CHAPTER ELEVEN

THE TELEOLOGY OF THE HYPERCONE

It will be seen how the regenerative concept of galactic evolution depends on the interlinking of quantum fields of vibrating energy, while the energy conversion processes which fuel this field relative to a matter creation/annihilation process dependent upon electromagnetic forces.

Magnetism on a cosmic scale is dependent upon the largescale movement of charged particles, which, it has been shown, relates quite exactly to a relativistic distortion of spacetime. It would seem therefore that gravity relates directly to magnetism as much as to the presence of matter. We see that objects with progressively stronger gravitational fields exhibit equally strong magnetic phenomena. The magnetic field of a white dwarf, for instance, is a billion times that of the earth. The difference is that there are practical limits to the intensity of magnetic fields which do not apply with mathematically based general relativity. One needs moving charged particles to create them, which cannot travel faster than light.

Being open-ended the hypercone type of universe offers solutions related to high translatory velocities of charged particles, giving rise to magnetic effects impossible in a universe of the hypersphere type. I shall develop this aspect in the second part of this book.

in the wider context, hyperconism raises a new and strange edifice which cannot but have repercussions upon the philosophy of science. In essence it is a more biological approach to astrophysics, since it affirms that nature usually works to set systems in motion which thereafter continue to function and evolve in their own peculiar ways.

From the time of Copernicus scientists seem to have been making discoveries which dismantled strongly held religious views and scriptural theories. So they seemed to be threatening the moral basis for human behaviour. Scientists may therefore have acquired a guilt complex about this, so much so that Einstein sought to establish a rule of law in the universe, rather than a basis of chance expectation. Meanwhile astrophysicists sought to restore some of the book of Genesis and the divinity of the circle.

Having removed the earth from the centre of the ptolemaic universe and given it no special place among a galaxy of stars in a vast, impersonal expanding cosmos, will this cosmos now begin to look more friendly if it is seen to be organically connected as part of a self-maintaining and self-reproducing system, a vast gaia complex of interlocking forces.

The existence of states outside the present moment of time has been hinted at by many theorists, not least the apostles of general relativity. What it seems impossible to come to terms with is that a physical world could be influenced by something pre-existing in the future, as this would violate all forms of causality. All I say is that the power of fatality only becomes important if we accept it. We must take arms against a sea of troubles and overthrow them.

Perhaps the most hopeful aspect of this new idea is that

of an improving principle which is inherent in a gravitational field, which in the final analysis pulls matter into an ordered, self-regenerating state. Though this is classed as a weak force it may operate in a hierarchic progression from high energy/low organisation to low energy/high organisation, that is, from objects like quasars to elliptical galaxies and beyond to other forms as yet unknown. Truly the weak shall inherit the cosmos.

Eternal Recurrence

Writing in the *New Scientist* under the heading of 'The Anthropic Universe', Martin Rees expresses the view that there are so many coincidences in nature and in the way the universe was formed that the emergence of life seems to have been exactly contrived by the hand of some guiding intelligence. For instance, the weakness of gravitational forces made for large stars with correspondingly long lives, so their light could warm planets long enough for the evolution of life to occur. Similarly the expansion of the universe occurred at a rate such that the galaxies could form out of the primeval gases – if everything had been going a little faster this would never have happened, and if it had been slower everything would have collapsed into a big lump.

The composition of matter and the proportions of the elements had to be determined all in a very short period of time during the big bang, and the energy levels in certain critical elements had to be just right in order that the heavier nuclei could be formed out of them by supernova explosions.

In scientific terms all this argues against the chance that these manifold systems could have all come together to start the process of living evolution 'just like that' (quoted in memory of Tommy Cooper, the comedian, who was always getting his tricks wrong). The argument for a guiding intelligence on the basis of current theory seems unassailable.

However, I hope to have shown that much less was left to chance in the creation scenario, and the guiding hand had much less to do. Nevertheless the element of chance was always there and I have a distinct feeling that this universe was not a 'one-shot trial'. It seems much more likely that we are looking at a polished performance, honed to near perfection by an almost endless series of false starts and replays, revised at each performance to ensure the delicate balance of forces in the universe.

One finds the idea of eternal recurrence in very early Hindu scriptures. It covers a great age or manvantura, the day of Brahma, which lasts 4 billion years. The German philosopher Nietzsche compressed this to a mere 50,000 years in his writings. The theme has been taken up by other writers like J.W. Dunne in *The Serial Universe*, and in the plays of J.B. Priestley. All these thinkers may have been influenced by the Christian doctrine of the resurrection. This is a hopeful doctrine which implies that things will be better the next time around.

If repeated waves of the material present do wash around the breakwaters of a cosmic energy field then it behoves everyone to try to do better each time they get the chance. In this way living will become more tolerable, the frontiers of evolution, material and social, can be pushed back even further. Likewise the performance of life on the great world stage will always take on new facets of interest and entrancement.

I now turn to a consideration of the microcosm in part two. In this study I hope to be able to show that the criteria of wave propagation determine that the separation of recurrent waves of physical universes could be about fifty million years. By this token one needs to acquire a great deal of skill to while away the time in the waiting room of the purgatorial station between the stars.

PART II

CHAPTER TWELVE

THE SEARCH FOR AN AETHER

At the outset of the summary of theories relating to the very small let me explain that 150 years ago God was in his heaven and all was right with science. There was a universal belief in the trinity of matter, waves and the luminiferous aether. This aether was the medium by which electromagnetic waves were carried. They were transverse waves it was true, but sound waves passing through a solid were transverse, so the aether must be something like a solid.

The relative speed of matter and waves could vary, but the speed of a wave relative to a homogeneous medium which carried it was always a constant. One has the simile of a ship sailing on a calm lake. The ship can change its speed, but the waves it makes travelling over the water have a constant speed relative to the water in the lake. The ship experiences a frictional drag in moving through the water and pushing aside the bow wave.

Physics tells us that waves travel faster as the medium becomes denser. Waves travelling through a solid are transverse, moving from side to side inside of producing longitudinal zones of compression and rarefaction, as sound waves do in air.

If I make a sound at one end of the desk I am writing at, the sound waves take about 3 milliseconds to travel to the other end in air. But the sound travelling through the solid wood goes twelve times as fast, and in glass seventeen times as fast, or 5,000 metres a second. But light travels at 300 million metres a second, therefore the medium that conveyed it must be very hard, absolutely rigid, frictionless and perfectly transparent. The magnus aether, indeed, was many times harder than the hardest crystal of diamond.

Small wonder therefore that there was a need to prove or disprove the aether theory. This would be done by somehow measuring the drag produced by the earth in its movement through absolute space, which would surely produce some difference in the velocity of light in different directions. Remember that in some ways our earth was seen as the ship sailing on the lake, and if the trinity of matter, waves and medium were true, then this ship could vary its speed relative to the waves.

So the test of the theory of the luminiferous aether was the famous optical interference experiment carried out by Michelson at Cleveland, USA in 1881, and repeated in 1887 with greater accuracy, assisted by Morley.

Interference of two trains of light waves occurs when they are superimposed. If the waves beat together in step they add up, if they are out of step they cancel each other leaving a dark space. In the Michelson-Morley experiment, beams of light were sent across the laboratory, then reflected from side to side, and then finally back along the original path, so that the two beams interfered. If there had been any difference in the speed of light in one direction or the other then this would have appeared in interference fringes, patches of light and dark. There was none. Though the experiment has often been repeated at different times

A Model for Wave Motion

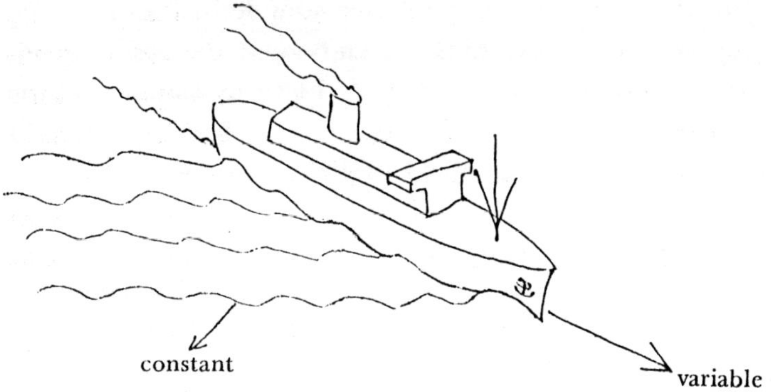

Michelson-Morley Experiment

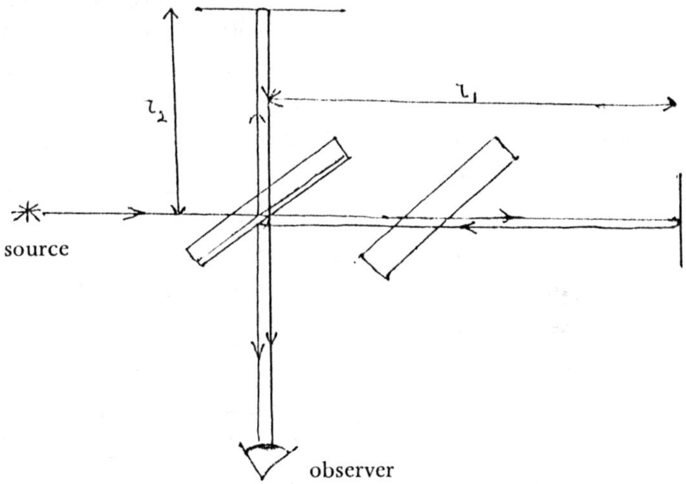

of the year and with more modern apparatus and using radar over astronomical distances the result is always null.

This experiment is said to have sounded the death knell of the aether theory. Einstein embodied the result in his theory of relativity in two ways. Firstly by saying that the laws of physics remained the same in all inertial frameworks and secondly that the speed of light was the same in all inertial frameworks.

One can understand the reluctance of scientists to believe in anything with the fantastic properties accredited to aether, and certainly, following the Michelson-Morley experiment, the trinity of physics had been reduced to a duality. It is easy to argue that one should chop off the unseen and incomprehensible entity of aether with Occam's razor, but this is rather like the fish in the pond saying: 'I don't see the water therefore I can do without it'.

Looking at the problem in terms of simple logic (which has got scientists into trouble on many occasions). One has the proposition 'Man is mortal', then a statement: 'Socrates is a man', therefore, 'Socrates is mortal'. So, by the same reasoning 'Waves have a constant velocity relative to their medium' is the proposition. The result of all the experiments show that light waves have a constant velocity relative to the earth. Ergo, one of them is the wave and the other the medium. This is simple logic and yet light is assumed already to be a wave and matter cannot be the medium for light, since light passes through a complete vacuum. Science could not accept the incredible conclusion that was left – that matter was the wave and light a manifestation of the medium through which matter moved.

I ask the reader to recall the analogy of the spaceship travelling at light speed in part one of this book. Imagine the whole of four-dimensional space is filled with an energy field, or quantum field if you like. The spaceship is reduced to a time cone of three dimensions. Along this cone, light signals are seen coming from all of the three dimensions. Beyond this thin slice of the present moment nothing can be seen. In this part of the book, I have replaced the spaceship with a boat crossing a lake. The whole of the lake is the energy field, a representation of four dimensions. To make the boat represent three dimensions in the same way it has to be squashed into a thin slice across its axis of motion, and the bow wave becomes likewise a projection of the time cone, the only place where one would see light in the analogy. In the Michelson-Morley experiment there was no way that they could position their apparatus along the earth's axis of motion, which is represented by the boat in this analogy. This is because that dimension had collapsed due to the earth's speed. They were left with the other three dimensions in which events arrived from all directions at the speed of light, and departed in a similar fashion. Light signals could not be seen outside the thin slice of the time cone, because of the earth's speed relative to the reference frame of the cosmos.

Looking at the diagram of the experiment you will see it was only performed in a plane of two dimensions. Had the vertical plane been included interference fringes would have been seen, due to the effect of gravitation on the energy of ascending and descending light rays. These effects would have been dismissed for this reason – yet, as I hope to show, gravity was the actual manifestation of aether drag they were looking for!

Returning now to the history of physics it took the brilliance of Einstein to cobble a theory which made matter behave like a wave, becoming dependent upon light signals, without actually saying that matter was a wave.

Anyway the luminiferous aether theory was dropped and we were left with a duality in physics, a duality of particles of matter and particles of radiation. In that same momentous decade of the 1880s Rontgen and Becquerel were discovering nuclear radiation and the immense power locked up in the atom. This lead Einstein in 1905 to propose his famous equation: energy = mass x the square of the velocity of light. So you can see what a vast amount of energy is released when matter is changed into radiation.

Let us compare Einstein's equation with the one for kinetic energy, which equates to half the mass times the square of velocity. This is the kind of energy possessed by a body in motion, like the pendulum at the bottom of its swing. As the bob of the pendulum swings upwards it slows as kinetic energy is converted into potential energy, which is equal to mgh, where g is the force of gravity and h is the height. If one stops the pendulum at the bottom of its swing then kinetic energy is converted into heat, or radiation.

Einstein's equation suggests forcibly that matter is something in motion, and that when we stop it changes into radiation and its substance vanishes.

If one really believed in matter as something solid wouldn't it have been more logical to imagine it contained potential energy – mgh? This potential energy would then have been converted into kinetic energy when matter became radiation, which spreads out in every direction at the speed of light.

To put this equation into the potential form would have created difficulty since radiation is weightless when at rest and the energy necessary to accelerate it to the velocity of light is only the energy of its vibration. If the problem is turned around, and matter becomes the moving wave, radiation the effect of an impact with a static energy medium, then the energy equation is at once clarified. But there is one snag, matter has to travel at the speed of light, for the constancy of the apparent velocity of radiation is relative between the wave and its medium.

I now return to the contention in part one of this book that the considerations of special relativity apply only to motion in the three physical dimensions. For instance, the fact that clocks slow down at high speeds is due to this secondary motion being deducted from the actual velocity of matter. Time progresses at c, the speed of light, and when matter attains this speed in one of the other three dimensions of space it is slowed to a stop. Secondly matter has inertial mass, electric charge, magnetic moment and gravitational attraction, possibly as a result of its motion at high speed, the static energy medium does not have these properties, motion through it could be assumed to produce only a linear variation of the properties of matter up to and even beyond the speed of light.

The equations of special relativity produce a limitation on the speed which can be attained by any material body, because it requires disproportionate energy to travel very fast,* and infinite energy to attain the speed of light. But as I have been hinting, this may only apply to a secondary velocity in a world of three dimensions, the diversion from a primary velocity of the speed of light through an energy field. True, that primary velocity has caused matter to

*$E = E_o/(1-V^2/C^2)^{1/2}$

collapse along the axis of its motion, but this is a linear effect because matter is a wave, and it does not mean that matter could not collapse further if its speed were greater. The speed of light at which it travels just happens to be the speed at which it was launched by the big bang, and the speed it maintains relative to the energy field.

Let us consider further the nature of light produced by the interaction of matter with the energy field. By the above reasoning radiation does not experience time progression, after being created or extinguished by contact with matter, it exists or is cancelled forever, unless matter contacts it again. Radiation is static and immutable, and within an energy field it is eternal.

To explain this idea it is necessary to continue with the historical sequence of events in scientific discovery. In 1905 the corpuscular concept of radiation was reinforced by the quantum theory of Max Planck. His idea was that radiation was emitted in packets, quanta or photons, multiples of a fundamental unit of energy, h or Planck's constant. So energy was given by h times ν (Nu), where the Greek letter Nu stood for frequency.

When one examines evidence for the translatory velocity of light can one distinguish whether the photons actually move or whether it is us that are moving and colliding with them? Cloud chamber photos can be produced which show particles being hit by gamma rays and breaking up, we can measure the effects of electrons being knocked out of their atomic orbits by radiation, when one switches on a light the heated filament appears to give off light. But in a way it's the same as being bombarded by high velocity shells, you don't see the one that hits you. You certainly cannot see a photon before it hits anything, this entity is chargeless and has a rest mass of zero, it doesn't leave tracks. One has to infer what has happened from the mess that it leaves around. In short, one cannot see a photon in motion.

The clincher is of course whether two photons ever collide. If they were static entities they couldn't. Energetic photons can produce pairs of particles by photoproduction by interaction with a field of matter. The phenomena of interference has already been mentioned which can produce light or dark patches where light adds together or cancels itself. This requires various material devices such as diffraction gratings, slits, mirrors, wave guides and aerials. But with the sensitive apparatus now available for the detection of radiation, it has been found that even a single photon will produce interference patterns when such devices are used. So one could say that the collision of two photons is not necessary to produce interference. Interference results from the use of specially-tuned, material devices. One can say the same of curved mirrors and lenses which concentrate light. Refraction and reflection of light can be seen as recreating photons in positions where we or other material objects will collide with them in our passage through time.

The question is a little confused because the electron behaves in some ways like a wave. It can produce characteristic diffraction patterns by interfering with itself. Electrons however, can be seen moving by the trails they leave as charged particles. Electrons also move in response to electromagnetic waves through conductors, and in this electronic form waves of radiation can be mixed to give higher and lower frequency harmonics. But this does not occur in vacuo. If it did the characteristic colours, or spectra, we receive from distant stars would be changed and unreliable. Doppler and gravitational shifts do occur in spectra due to velocities of approach or recession and intense gravitational fields, but in these cases the pattern of colours moves as a whole, so that the eternal integrity of photons cannot be doubted.*

Left to themselves in space photons are never seen to interfere with each other or collide. If they did they might aggregate into dangerous bundles of radiation. But they always seem to arrive as they set off, unless matter gets in the way.

So we come in a roundabout way to this peculiar business of how light manages to travel in straight lines, having no rest mass. If instead matter has such a high velocity that it can travel in a straight line through an energy field, then it is just an appearance that a static photon whizzes by in an apparently straight line in the opposite direction. This simple transposition of effect is clearly not the whole story because it does not define the nature of a photon, how it comes to be immutable over great distances. This is all connected with the properties of the energy field, which I shall try to relate to similar properties of material mechanics.

CHAPTER THIRTEEN

A FUTURE CONFIGURATION OF EVENTS

To continue with the history of science, in the 1920s a Frenchman called De Broglie looked at the two equations:

$$E = m.c^2 \quad \text{and} \quad E = h.\nu$$

And since they were both equal to energy he made a third equation out of them, replacing frequency ν with c/λ, where λ is the Compton wavelength of a material, so he got:

$$\lambda = h/m.c$$

In his thesis De Broglie chose his words carefully, saying only that there appeared to be a wave associated with a particle. The shortest wavelength possible was attained when the particle was travelling at the speed of light. This just happened to be the Compton wavelength, or the apparent size of a particle when it collided with other particles, or its collision cross-section.

De Broglie therefore enunciated a linear contraction of wavelength corresponding to velocity which was quite different to the Fitzgerald contraction used in special relativity, which squashed objects along their line of motion until they were reduced to a surface vertical to that axis at the speed of light. Special relativity introduced an asymptote, or limit, of light speed, which it was impossible to exceed. De Broglie's equation, being linear, merely stopped at the speed of light, but did not exclude the particle from going faster. It stopped at the speed of light because that was the Compton wavelength, and one is left with the surmise that this might be because a particle actually transits the energy field at this speed. De Broglie's equation thus can be seen to define a linear contraction of length as one goes faster through an energy field, not a Fitzgerald type of contraction. This again suggests that the energy field is not limited to the speed of light.

I have been talking glibly about transiting an energy field as if it were a material continuum. But in my thesis this energy field is related to time and there is apparently no translatory movement in the three-dimensional world which corresponds to motion in time, except at immense distances, related to Hubble's law. I have suggested that the fourth dimension of time has been collapsed by the speed we pass through it, so that along a time axis we only perceive a fleeting moment. Now I have added the conclusions that the flattening is linear, and the confinement of particles to our present moment is related simply to their speed. I have also suggested that the particle has acquired a species of kinetic energy and other properties resulting from that speed, but could all this result instead from an energy flux, let me explain further.

One has to remember the arguments Einstein used to abolish the idea of gravitational force in general relativity. The question is whether a force can be removed if you accelerate your co-ordinates instead, moving space relative

Time Cone

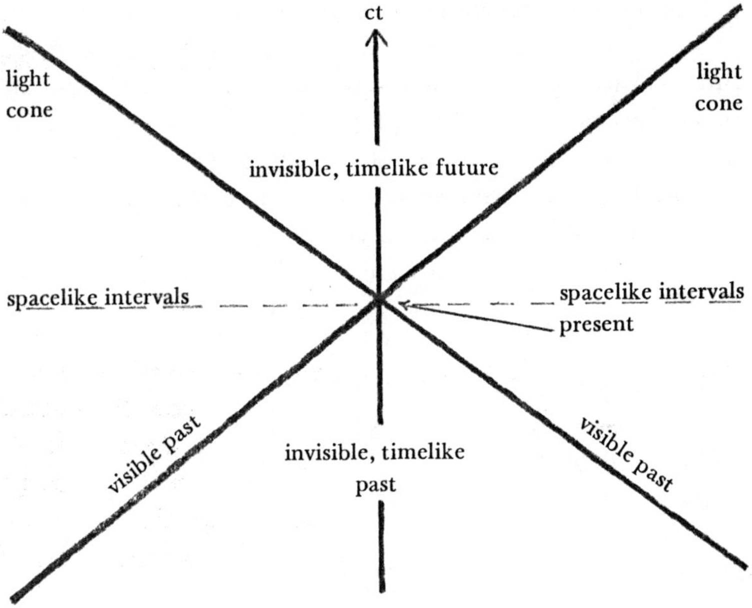

Wave Group

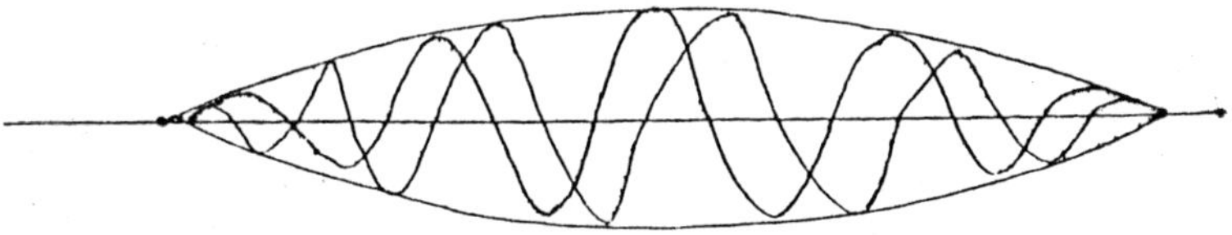

to yourself so that the force is cancelled. This is like shifting the goalposts down the pitch to where you are so you can score while standing still. Is there an equivalent method with relation to an energy field by which we can abolish the actual physical movement of all particles at the speed of light? This analogy for movement in a energy field, or time continuum, is rather like the de-tuning of a radio signal received at one point on the frequency scale of a radio set. When one moves the tuning knob of the set the station being received seems to become more distant, as though it were receding from you, while the station has not in fact moved. What you have done is to produce a change, or flux, in the received energy within a broadcast range of energies.

If the tuning dial is moved slowly this would be comparable to the translation of our material universe through the energy field of time. The broadcast signals of many stations exist at all times, and this epitomises the eternal quality of radiation, but we can only pick them up as we reach that point on the tuning dial. Continuing this analogy, it would be as if we were only able to turn the dial in one direction, following a process of entropy and the arrow of time.

How can we verify that a time continuum exists? Well, in every direction that we look we see events down the sides of a time cone, as they were a certain time ago depending upon the time it takes for light signals to reach us. But there is also a 'present moment' for these distant places, which is timelike and inaccessible to us, displaced ahead of their position along the time cone. In this way we are aware of an extension through time.

Returning to the question of what a photon is, we are

all familiar with the idea of radiation, or waves spreading outwards from a point of disturbance in an expanding sphere in space. Because of the speed at which light travels the radial dimension of the wave becomes collapsed and the wave only exists in two dimensions. The receiver of the radiation contacts the expanding sphere at one point with the same relative velocity, producing a further dimensional contraction to one dimension, or a line. One now has a possible reason why light is observed to move in straight lines – this is because it can only exist for us in one dimension, a line drawn from our point of contact with the wave back to the origin. The line, if drawn through an energy field, would contact points of comparable energy in that field in terms of median line, amplitude and phase. Contact with any point on the line has to draw the whole of its energy, instantly.

Here one meets a difficulty with this theory because a radiating source has to be drawing these lines through the energy field instantly if they can be instantly cancelled. Although I have hinted that something may transit an energy field faster than light-speed, the idea of instant creation and destruction of a line through the whole of the cosmos is something hard to swallow. What I hope to show is that communication of this kind does occur and to postulate a theory as to why it happens.

Firstly one has the picture of an energy field sloping downwards and outwards in all directions of three-dimensional space from the instant of the big bang. This determines the direction of the arrow of time. But within this four-dimensional continuum there are three-dimensional zones of equal energy, what distinguishes them? The answer must be the three-dimensional phase state of the energy field. Within the extremes of the field are phase states differing by 180°. A transfer of position within any point of a three-dimensional field would thus only depend on an oscillation through 360°.

It is an accepted dictum of science that phase states can be transmitted at a greater speed that that of light, since they only convey information. In any group of waves the speed, w, at which pilot waves, or changes in phase, can be transmitted through the group, is given by:

$$w = c^2/v \qquad \text{(where v is the velocity of the group).}$$

This shows that if the velocity of the group falls to zero the speed of transmission of the phase state becomes infinity. It is therefore necessary to postulate that the group velocity of a photon is zero. It can then circumnavigate the cosmos in an eyeblink.

This is not likely to be a theory which would commend itself to science, since on reflection one can see it implies a future configuration of events. What happens now when one sends out light signals is that one draws instant lines through space, which later on the receiving station will contact by moving into a condition of the energy field which matches that of the photon.

However, causality within our material world of three dimensions is preserved. Science already accepts that large astronomical events, like a supernova, send out light signals which travel through space for millions of years before we see the event. In a sense our future reception of the light energy is pre-determined the moment it sets off. All the present theory says is that the future has happened here before we perceive it. It has happened in a part of the energy field through which we are going to travel. Science already has a theory about tachyons, which are said to travel faster than light. Such tachyons if they existed, would give us a warning of a supernova event in the same way.

CHAPTER FOURTEEN

VORTICES IN SPACE

Having defined a photon as a species of instant line created through space the reader may be wondering how this fits in with Maxwell's equations which defines the transmission of electromagnetic energy through space. I shall now endeavour to describe these equations in the context of what I have been theorising.

The student is first taught Fleming's right hand rule for obtaining the direction of electric and magnetic force around a conductor. If the first finger points in the direction that current flows from the positive terminal of a battery along a conductor, then pointing the thumb outwards at 90° gives the direction of magnetic force, while bending the second finger across the palm gives the direction of the electric force. As current flows in the conductor the hand must turn to the right to follow the movement of the field.

Maxwell's equations transpose the electromagnetic field into space. There is still a twist to the field as the wave progresses forward, so thumb and second finger, spaced out from each other at right angles, and at right angles to the axis of motion, describe spiral paths, creating a double helix. This twisting around the axis of motion is called mathematically a curl of a vector. The curl of the electric field, or curl E equals:

$$\text{curl } E = -\partial \quad B/\partial t; \text{ where } B = \mu H \text{ in gauss}$$

The curl of the magnetic field equals:

$$\text{curl } H = \quad J + \partial D/ \quad t; \text{ where } D = \quad \mathcal{E}\, E \text{ and } J \text{ is potential}$$

The twist of the spiral creates a divergence effect, called div. The spacing of divergence and convergence of the spiral creates a wavelength.

$$\text{div } B = 0; \text{ div } D = p$$

Where magnetic or electric poles exist there is a slope, or gradient to the field, called grad.

The flow lines of current form closed paths according to Maxwell. This means that they must pass through the insulating dialectric of a capacitor, and also through any vacuum between the plates of a capacitor in a vacuum. This implies that energy can be stored in the medium of empty space by elastic distortion. Since the empty space surrounding a conductor is capable of this elastic displacement it is also able to transmit energy, and so displaces an equivalent current in any closed conducting loop in its path. This reversible process forms the basis of wireless communication.

Maxwell imagined that space contained imaginary cells, or vortices, which served to twist the field around a conductor as the current flowed along it. The process is reversible, if one instead moves the conductor through space in a surrounding field, then current flows through the conductor. Relatively nothing has changed. So, by analogy, making a particle move at the velocity of light through an energy field is simply a reversal of the same character – the exception being that this system allows one to account for particle spin more easily, as this would arise from the vortex nature of space. One could also postulate that particles behave like current loops and interacted with magnetic fields to draw energy from space. Spinning such a collection of matter consisting of current loops would then begin to explain the behaviour of gyroscopes, and angular momentum.

The same argument might be used to explain the mysterious integral energy orbits which electrons weave around the atomic nucleus. These could be constrained by the vortices of the radiation field. This question has been highlighted recently in a paper by Harold Puthoff (*Physics Review* D, Vol. 35, p. 3266), explaining why electrons don't as a rule fall into the nucleus. On the rare occasions that this happens, it is deemed to be a form of radioactive decay called 'K electron capture'. Puthoff's explanation is that there exists everywhere an untapped potential energy, or zero point energy. Normally this potential energy is balanced, but its presence is revealed by the Casimir effect of force between two metal plates in a vacuum, when the balance is somehow disturbed. Puthoff theorises that electrons in the lowest energy, or ground state, orbiting around the nucleus of the atom, are able to continually absorb energy from the zero point background, this underlying sea of electromagnetic energy. The way in which the absorption occurs can be more easily appreciated if it is necessary for electrons to follow certain well-defined paths through spatial vortices.

Maxwell's equations were formulated long before the quantum theory and relativity. His ideas of space gave credence to the idea of a luminiferous aether. But although the aether has been dismissed, the equations have a new relevance in respect of the quantum field. Moreover, in relation to the ideas I am now putting forward, the individual vortices which Maxwell proposed can be seen as photon lines. The energy of the photon line can be seen as due to the rate of flux of the electric and magnetic components, so that the energy steps of h, or Planck's constant, must be related to spatial conditions which preclude any intermediate variation.

The existence of the quantum of action is crucial to wave mechanics theory and its value was inferred from the curves of black body radiation at various temperatures. The value of h is calculated therefore from a relationship between energy and frequency, a wave of a certain energy, corresponding to a certain temperature, must be a given multiple of h. The value of h is thus derived empirically and cannot be deduced from any independent formulation of the constants of nature. This is different from c, the speed of light, which can be measured empirically as the time taken by light to pass between two points, or alternatively it is given by the formula:

$c = 1/(\mu \cdot \mathcal{E})^{1/2}$; where μ and $\mathcal{E}$ are constants representing the permeability and permittivity of free space.

The above constants are so contrived as to complement each other in a system of rationalised units. They also relate directly to Clerk Maxwell's equations for electromagnetic waves, given above, such that the vector product of curl H and curl E yields a second order differential equation with an oscillatory solution as the basis for the harmonic motion of waves in space. This vector product has the dimensions L $.T^{-2}$, wave energy having no mass. To make these equations relate to mechanical energy one has first to divide by $1/c^2$ (since $E = m.c^2$).

No possible limitation of the time element could have been foreseen by Clerk Maxwell, since it is only in this century that we have become aware of the expansion of the universe, which puts a time limit on its existence. This limit can be set at minimally 2×10^{17} seconds (7 thousand million years), although current estimates are three times as high.

Planck's quantum of action h has a value of 6.623×10^{-34} joules/seconds, which has the dimensions of $M.L^2.T^{-3}$. To relate this to the spinning of a particle it is usually divided by 4π. The following relationship can thus be proposed:

$$h/4\pi \sim \text{curl } H \times \text{curl } E/c^2 \; .t;$$ where t is the age of the universe.

The properties of waves in free space can thus be related in a classical sense to quantum mechanics. The rotation of Clerk Maxwell's wave vectors as a system of helices in space which rotate both the E and H fields following the direction of the wave in space can thus be seen to influence the quantisation of particle spin in precise units of $h/4\pi$, as has been revealed by the precise separation of spectral lines in a magnetic field.

This relationship has an elegant simplicity which should appeal to purists.

It will be seen that there are wider implications to such a relationship. Firstly, it is proposed that the value of h is determined by the extent of the universe. Secondly, that the association of the value of h with spin energy points to the conclusion that spin is inherent in the structure of the universe. Such is the case with the hyperconic form already suggested.

It follows from the above that the value of h cannot be a constant, its value will fall in time. Given that the total energy of the universe does not alter, then the frequencies of photons should increase with time, since it takes less energy to create the same photon at a later time, since $E = h.\nu$. This, if you like, is a transposition of Hubble's law. It has the advantage that it does not require the universe to be homogeneous and isotropic, a doctrine which is under threat anyway.

In addition to the reasons already advanced for such an approach it is clear that, in an expanding universe, there is reduced energy density as time passes. Such reduced energy density is another reason why one should expect a lowered value for h and a greater freedom for particles to oscillate with less expenditure of effort.

The reader will be aware that cosmological models which rely on the least a priori assumptions have the greatest validity. I have advanced only one general postulate so far. This is, that the universe was launched into existence by a high energy impulse. I do not even require that impulse to have been uniformly distributed in all directions. It follows from this that the high level of initial energy declines as time passes, marking out various epochs of development, and that the gradient of energy decrease is associated with a pseudo velocity c, the speed of light.

Maxwell's wave equations prescribe motion in the form of a helix around the axis along which an electromagnetic wave is transmitted. By transposing the roles of matter and waves, so that matter is the wave travelling at the pseudo velocity c, I am saying that space is filled with photon helices which matter must traverse. These helices impose rotation on particles and this solves the problem of angular momentum which bedevilled both Newton and Einstein. The existence of spatial helices also explains why current carrying coils tend to bend into a circular form when moved at high velocity through a magnetic field, thus creating closed current loops as the basis of particles and thir fields as will be explained later.

There is a last consequence of this series of transpositions, which has already been anticipated by the technology of dating rocks and fossils by their radioactive signatures. It would no longer be appropriate to specify time as 't' in any equation. Time is now linked with an energy/frequency state and belongs to an epoch whose precise conditions will not recur.

It is customary to suggest tests for every new scientific theory. If this theory is correct, then every 32 years or thereabouts, the value of Planck's constant h will decrease by one part in 600 million. The astute reader will realise that I am not likely to be proved wrong.

CHAPTER FIFTEEN

PHOTONS AS CAUSAL CHAINS

From what has already been suggested one has the picture of a photon consisting of a double helix extending through space, brought into being instantly at the moment of its creation. If a particle contacts this helix subsequently, when both happen to share exactly the same energy conditions, then the helix disappears along its entire length. The energy conditions of the photon do not change so long as it remains in being, so, in effect, it is lying in wait for the particle to come along, in the spacelike region of the universe.

One has the additional presumption that a close encounter with a particle will change the direction of the photon helix, without causing it to collapse. This refers to interference experiments, where light is passed through double slits separated by a fraction of the wavelength of the light. A screen placed beyond the slits will then show characteristic light and dark bands, which are the aggregation of numerous impacts of photons, where in some places the energy is added, and in others, due to the photons being out of phase, the energy is cancelled.

If the interference slits are opened and shut alternately the interference pattern on the screen is destroyed. These effects persist although only one photon may pass through the slits at a time. The single photon seems to experience a random deflection, even though it may not be able to pass through both slits, and successive photons gradually build up the characteristic diffraction pattern.

Though the problem of how a single photon can interfere with itself, or by chance alone build up a diffraction pattern, may seem intractable, nevertheless such behaviour can be dealt with by the laws of probability and chance expectation which go to form quantum theory.

What I would like to propose however, is that this is not the end of the story. The laws of chance themselves must be derived from some other system of deterministic action which has so far escaped notice *because it happens too fast.*

The peculiarities of quantum mechanics have attracted much attention in recent years, and it is customary to ascribe such peculiarities to 'hidden variables' in the experiment. What I would like to do instead is to try to formulate a logical hyperstructure to the material experiment which would then determine the chance outcome.

I have suggested that space may be riddled with photon helices, brought into existence on the instant by particle collisions or transitions. A photon helix thus created is at the apex of the time cone in its locality and thus has freedom to range the cosmos in the spacelike dimension. In the interference experiment therefore, the photon source lays down a photon helix in this spacelike dimension before the two slits arrive from further down the side of the cone. The photon helix is thus in a sense already beyond the slits when they arrive in the present. The arrival of the slits changes the condition of the helix without entirely cancelling it. The photon's helix cannot be divided but it can be modulated, therefore its incident energy will depend upon

Interference of Light Waves

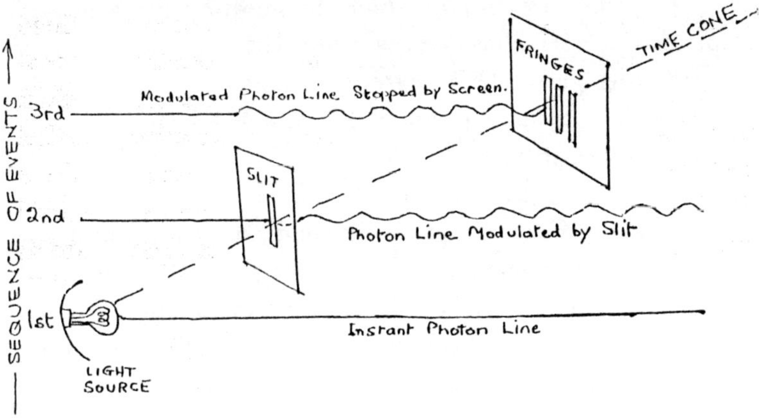

Polarisation of Light

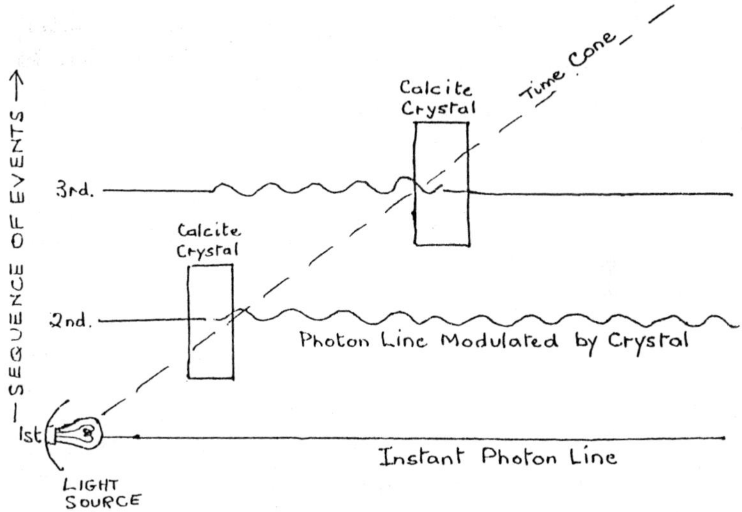

the phase of the modulation. Thus it is possible for interference to occur with just one photon.

The precise direction in which the photon helix is moved depends upon its phase at the point where interaction with the slit took place. This phase is changing through 360° so subsequent photons will gradually fill out a complete cycle of interference fringes.

What interests me particularly are the dark zones where two photon helices, or one helix changed into a modulated replica of itself, manages to cancel itself out by virtue of being in phase or out of phase with the modulation on impact. Were the helices waves, this would present no problem, but I am suggesting that the photon helices and their modulations are distortions of space time. Looking at it another way, I have converted particles to a wavelike role. If a particle and its anti-particle meet then they annihilate each other and radiation appears in their place. Likewise an energetic photon can produce a pair of particles by interaction with matter. In no case does energy disappear. So what happens to the energy of a modulated photon vortex? The solution I favour, and which will be developed in this book, is that the cancelled energy serves to displace the zero energy point of the quantum field, which means that these photons have to appear at some other point in space-time.

Before leaving the subject of interference however, there is another important experiment which is worthy of mention. When light is passed through a calcite crystal then there are two emergent beams which are polarised parallel and perpendicular to a particular crystal plane. If these divided beams are subsequently passed through a second crystal whose planes are reversed, then the light beam is recombined and it would have the same polarisation as the incident beam.

When the light beam used in this experiment is so weak that only one photon can pass at a time, it might be thought that the initial polarisation state the photon had might be lost when passing through the two crystals, since one could argue that a single photon must follow only one of the two paths in the split phase, and so would not reunite with anything that could set the polarisation back to what it had been before. This is not the case however, the single photon will always emerge from the second crystal with the same polarisation intact, just as if it had been split between the two crystals. This can be further tested by blocking off one of the intermediate paths, when indeed the photon loses its polarisation state.

Though it is said the photon only goes by one path it seems to know what should happen if it went along the other as well and it retains a memory of this condition to enable it to resume its first state. To account for such occurrences in a rational way the hidden variable theory suggests that perhaps the path of the photon has some hidden property *which determines its behaviour in advance.*

One now only has to substitute the idea of the photon helix which pre-exists in the spacelike dimension along the path the crystals and the detector will occupy in an instant of time. The first crystal will influence the photon so that its energy is modulated, the second crystal removes the modulation. The modulation of the photon effectively broadens its path, and if part of this path is obstructed then the de-modulation by the second crystal will not be exact.

It will be seen that this solution does not pre-empt causality, the result of the experiment is simply not dependent upon the speed of light, since it involves paths existing in a spacelike dimension.

The reader may feel that I am being a little conservative about the principle of conservation of energy. No more so than Wolfgang Pauli when, in 1936, he postulated the existence of a new particle, the neutrino, which was necessary to carry off a variable amount of momentum in the beta decay of the neutron. We should be equally jealous of this principle – that energy cannot be created or destroyed. Energy can be transmitted or transformed, but the idea of cancelling it, or conjuring it out of the void must be viewed with suspicion.

CHAPTER SIXTEEN

A BRUSH WITH NON-LOCALITY

Non-locality is a comparatively new buzz word in physics, and to understand what it means one must go back to 1935 when Einstein, Podolski and Rosen were able to show that quantum theory implied that separated particles could influence each other although there was no apparent means of communication between them. This was an attempt to discredit quantum mechanics, which Einstein did not like. In 1951 Bohm crystallised this contention of the worthy trio, now familiarly reduced to their initials as the EPR rule. He did it in the particular context of photon polarisation, which is related to the spin of the emitting particle.

In certain circumstances atoms in an excited state can return to the ground, or low energy, state by the emission of two photons in quick succession. When this happens one photon will be polarised at + 45° to the horizontal and the other at −45°. Not all atoms which emit photon pairs have this property prescribed by quantum theory, but a calcium vapour light source does, and was suggested for the EPR experiment by David Bohm. Using calcium vapour in the experiment, one emitted photon is of green light and the other is blue and this fact can be ascertained by passing the photons through appropriate filters which accept the different wavelengths.

As the photons move away from each other in opposite directions, separating faster than light speed, their polarisations can be detected by crystals of blue and green calcite and are found to be out of phase by the amount required by the theory. However, it is a peculiarity of quantum physics that the state of polarisation cannot be known until it is measured. The Copenhagen interpretation of the theory asserts that the act of measurement must change the state of the photon, when its probablistic state collapsed. If then, there is a synchronism between the states of the separated photons, as there appears to be, it implies communication between them, although they are moving apart at twice the speed of light!

This, in essence, is the problem of non-locality, which has been further reinforced by Bell's theorem, as propounded by John S. Bell in 1969, which shows that no hidden variable theory which preserves locality and determinism is capable of reproducing the predictions of quantum physics for the two-photon experiment. This means in simpler terms that if the hidden variable exists it has to be outside timelike, or normal, space-time.

The theorem itself relates to the chance expectation for three variables, each with two states, + or -, like the phase states of photons when emitted and when subsequently detected by the two calcite crystals.

Bell's theorem, sometimes known as Bell's inequality, is usually written in the form:

$$n(V, \phi +) + n(\phi +, \theta +) \geqslant n(V, \theta +)$$

Putting this into non-mathematical terms it states that, in any three experiments carried out with the apparatus mentioned above, where two photons are emitted at once and detected remotely, with the three following alignments of the detectors:

1. One vertical and one at an angle of θ to the horizontal;
2. One vertical and one at an angle of ϕ to the horizontal;
3. The right one on the left at an angle of ϕ and the one on the right at an angle of θ.

Then the total numbers of pairs in which both photons are registered positive in the first experiment can never be greater than the sum of the doubly positive pairs in the other two experiments. This is if the results are to be affected by hidden variables within our space-time and neither photon can influence the state of the other.

All that now remained was to prove that this inequality was false for some values of θ and ϕ. Though the Bell theorem is obeyed for values of θ greater than 30°, it is breached if it is between 0° and 30°.*

Numbers of experiments have been carried out over the past fifteen years to test Bell's theorem and these have all demonstrated the necessity for non-local connections between the emitted photons in order to explain their phase relationship. The most important of these experiments was by Alain Aspect in 1982, which supported the quantum theory and ruled out the Bell inequality, with its local and deterministic hidden variable. One is thus left with the EPR paradox of a faster than light system of communication between quantum states of photons which defies special relativity and is inaccessible to us in the material world.†

One thus has support for the idea of a photon helix

Alain Aspects Non-locality Experiment
(simultaneous events co-exist throughout space)

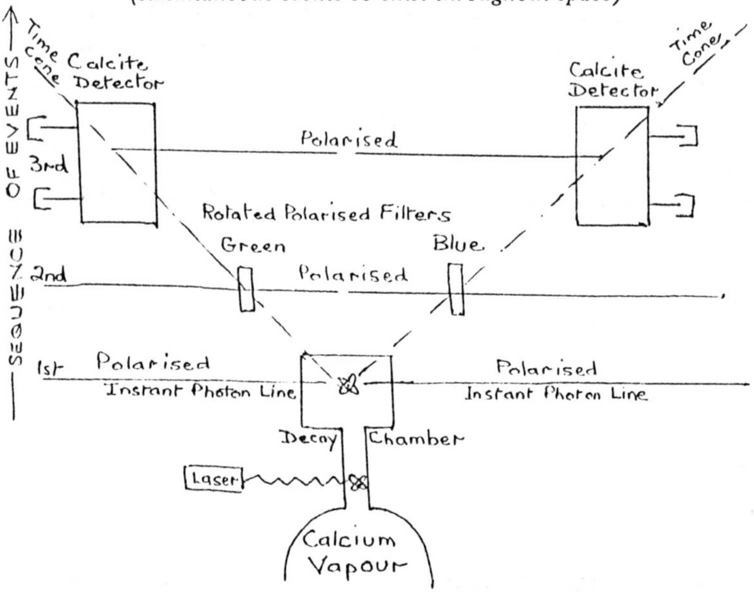

The test of John Bell's theorem
The experiment by A. Aspect, J. Dalibard & G. Roger

Atoms of mercury or calcium vapour travel upwards through a tube where they are hit by a laser beam, causing these atoms to absorb energy. They then pass upwards into the decay chamber and revert to the lower energy state by emitting 2 photons, which can then be described by a single quantum state (being simultaneously emitted). Two detectors record whether the polarisation of each quanta was the same, which allows them to pass through the polarisers. The angle of the polarisers was chosen at random after the photons were emitted. The theory of relativity does not allow them to communicate when separated, as they are travelling apart faster than light, yet their state of polarisation remains the same. This can be explained by instant transmission of a phase state through the scalar energy field, outside the timelike hypercone of real space, which is shown dotted on the diagram. The two, instant photon lines then determine a causal chain of events as they happen in sequence.

being instantly created in a spacelike dimension the moment the two photons are emitted from the calcium vapour in opposite directions. The blue and green filters next arrive on the scene from further down the slope of the time cone, and finally the calcite detectors, which register what the phase angles of the photon helices were when they were emitted.

The causal chain in the experiment is thus created in an instant throughout a spacelike dimension at a speed greater than that of light. Subsequently the causal chain of the photon does not move, it is the experiment as a whole which has to move up through time to give the illusion of light waves, or photons going from the source to the detectors.

What does quantum theory have to say about such an idea? Well, in the first place it would not presume to know anything between the time the photons were emitted and when they were both measured. The measurement is the thing that collapses the quantum state and makes probability a certainty. Whatever happens is reduced to the experience of the observer at the end of the day.

Secondly, quantum theory implies speeds greater than that of light, if the thing that moves is not a material body, but perhaps information. All of a sudden, when the observation is made, something seems to spread out in all directions instantly to encompass such things as the phase of a wave or photons at distant points.

The third point is that quantum mechanics does not see the universe as a number of separate and discrete atoms, like the old, billiard ball concept, but as a single, connected whole.*

Therefore I have some leeway in theorising as above,

that there is a spacelike dimension outside time, occupied by an energy field capable of transmitting information instantly, just as if it were a mirror that suddenly cracked from side to side. Perhaps the definition of a spacelike dimension outside time is too laboured – should I not call it eternity?

CHAPTER SEVENTEEN

SCALARS AND THE QUANTUM FIELD

A quantum field has been postulated as a background vibration which exists everywhere. It is likened to the pressure waves of sound. I have suggested that it supports the existence of particle and radiation, it could also maintain differences of electric potential, magnetic fields, even gravitational potential. This background field has been identified with a grand unifying factor in all physical theories.

The idea of a quantum field arose from the phenomena of photoproduction, already mentioned, whereby a powerful photon can interact with a field of matter to produce a pair of particles. Dirac postulated that in order for this to happen empty space must consist of an infinite sea of particles in negative energy states. This must be so for two reasons. Firstly, a negative energy particle must be present wherever photoproduction occurs, in order to be able to jump into existence as a particle pair. The particle represents about half the energy of the photon, while the other half is necessary to lift the particle from its negative energy condition. The hole that is left behind in the energy field represents the anti-particle.

Secondly, every available negative energy state must be filled by a particle to prevent real particles falling down into them. A filled state presents a barrier, because the Pauli exclusion principle states that no two particles can occupy the same energy state.

Prior to Dirac's theory, energy was not regarded as a scalar. A scalar quantity is one which can be transposed around a zero point, remaining in essence the same quantity. This is like a voltage which has the same electromotive force whether it is plus a volt or minus a volt, only the direction of the current is changed.

Energy was not a scalar because kinetic energy, as you remember, contains a square of velocity, and a square of a quantity cannot be negative. To have minus energy one needs to include an imaginary number, like the square root of minus one. Negative energy is therefore an imaginary quantity, something that does not belong to our timelike space. Making energy a scalar is convenient for quantum mechanics, which can deal with imaginary, or complex quantities. But if we are going to grasp the infrastructure behind energy fields we might at last be able to dispense with this kind of mathematical symbolism.

The further difficulty that arises with Dirac's theory is that all space is filled with this imaginary, negative energy. Such energy has a mass equivalent. This mass would be so large that space would tend to be rolled up by an infinite gravitational attraction. To avoid this a normalising factor has to be included in equations, which consists of another infinite force pulling against the first one. This is an unconvincing contrivance. One infinite quantity is enough bad news for mathematicians. Two are a disaster.

For the moment then, it is convenient to allow that there can be a scalar energy field, which has a zero point. What I have already hinted is that the zero point varies everywhere throughout the universe of three dimensions, it is part of the general gradient which determines the arrow of time. In an energy field which is a background of waves, position can then be linked to phase states, which as we have seen in connection with the Bell theorem, can travel instantly anywhere.

Let me come clean then and try to define classical mechanics in terms of what we already know about this energy field.

Quantity	*Unit*	*Dimensions*	*Wave mechanics*	*Symbol*
energy	joules	$M.L^2./T^2$	wave frequency	ν, ($h.\nu = E$)
force	newtons	$M.L/T^2$	wave amplitude	Λ
momentum		M.L/T	wavelength	λ
time	seconds	T	change of frequency ($d\nu 1$ at origin; $d\nu 2$ elsewhere)	$d.\nu$
position	x,y,z	L	given by 2 phase states and the elapsed time $d\nu 1$ at wave velocity c, the speed of light (i.e. polar co-ordinates).	$\phi 1, \phi 2$ & $d\nu 1.c$
speed	x',y',z'	L/T		$d/d\nu\ 2 \times \phi 1, 2$ & $d\nu 1.c$
acceleration	x'',y'',z''	L/T^2		$d^2/d\ (\nu 2 \times \phi 1, \phi 2$ & $d\nu 1.c)$
power	watts	$M.L^2/T^3$		$d/d\nu 2.\phi 1, \phi 2.\ A.\ d\nu 1.c$

In wave mechanics it is not convenient to use cartesian co-ordinates. Instead it is easier if position is defined from where you are, at your own time, by polar co-ordinates. is like aiming a telescope at a star, when one must know two angles, longitude and latitude of the celestial globe. The distance of the star is then the time that elapses for

for a light signal to travel between the star and the earth, sometimes many years. In wave mechanics these two angles are derived from the phase states of two wave components (like the photons in Bell's theorem) and the distance by the elapsed time at the wave velocity, or c.

Speed involves measuring elapsed time at a remote point which is different from your personal time, due to the fact that time varies from point to point down a time cone, and because the zero point energy of the scalar energy field varies in this theory.

The whole thrust of my argument therefore turns on the question of what happens when the zero point of the scalar field is altered locally. By my reasoning one would end up somewhere else, like the cancelled photons in the interference fringes.

The consequences of staying put while the zero energy point was shifted would be quite dramatic. By Dirac's reasoning, if the zero point was raised a lot of particles in low energy states, like tightly-bound electrons in atoms, would vanish. Similarly if the energy point was lowered, numbers of particles would appear out of nothing. In either case a nasty electrical explosion would occur – something like a bolt of lightning.

This is a good reason why we should be careful about making it happen. we could nevertheless use the displacement of zero energy in a scalar energy field as a means of transferring something like information, that is, as a very secure means of communication capable of sending messages instantly to any part of the cosmos. The information would not be broadcast, like a normal transmission, but would arrive in one lump as it set off. All the transmitter has to do, it seems, is to ensure that two wave components, having directional phase angles pointed at the place one wishes to go, and being each mutually cancelled by interference. I shall discuss the practical details of transmitting and receiving devices in the appendix.

To further reinforce my ideas on interference of waves one might remember that the amplitude of all oscillations are made up of sine and cosine functions. The energies of waves consist of the squares of the same functions, which are always positive. Now interfering waves give $\sin^2 \omega .t + \cos^2 \omega .t$ which equals 1. Therefore energy is not cancelled, but the zero point of the summed waves has been raised.

A background scalar energy field may also have storage potential for resonant vibration. All solid structures and particles have natural frequencies and modes of vibration. If such structures pick up mechanical, electric or magnetic vibrations then they can vibrate in sympathy if the vibration corresponds to their natural frequency. This is called resonance. Resonance consists of an amplification of the vibration in the recipient structure by the addition of successive impulses, up to the point where further increase of the amplitude of vibration is matched by forces resisting the excursions back and forth. If these forces are not strong enough then the structure breaks apart.

The factor of resonance and ability to store amounts of vibrational energy in nuclear and particle systems is implied by the excited states of nuclei, baryons and leptons. Synchronous release of such energy is described as laser or maser action.

A background quantum field may also have storage potential for resonant vibrations and this may offer clues to the nature of both strong and weak reactions, involving nuclear and gravitational forces. We have seen how oscillations can occur in free space in a perfect vacuum. This duality of electromagnetic waves which has been mentioned in connection with Maxwell's equations suggests forcibly that a scalar field supporting this activity would be dual in its nature, having two separate aspects which appeared in quadrature.

The magnetic force in a current carrier is smaller than the electric force by a factor $u.v/c^2$, where u and v are the relative velocities of interacting charged particles. The magnetic effects of the field can be amplified in the range of 1 to 100,000 by magnetic materials, while the effects of current in conductors due to potential differences are in the much larger range of 1 to 10^{20}. Direct electric current tends to confine itself to conductors. Alternating current and magnetic force, whether constant or changing, tends to spread ubiquitously through non-conductors, such as electrolytes, insulators or free space. Magnetism is defined as an effect due to the movement of charge carriers, which thereby produces a relativistic contraction of space. The fact that empty space can be modified in such a way as to produce a force though seeming to support the ideas of general relativity, is more probably the best argument there is to substantiate the existence of a scalar field existing in a vacuum.

Dirac's ideas about an infinite sea of negative energy states is actually bipolar. The field of positive energy particles stretches upwards from the level of rest mass energy of a particle. Below zero there is a corresponding field which stretches downwards from minus the same rest mass energy. The scalar energy field is thus bipolar, as might have been suspected. In the two diagrams I have shown

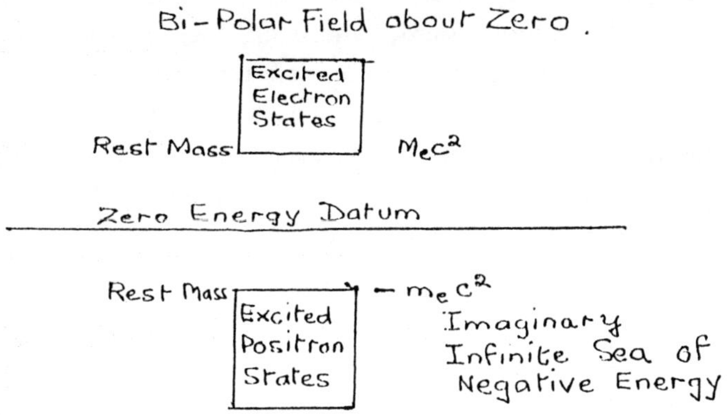

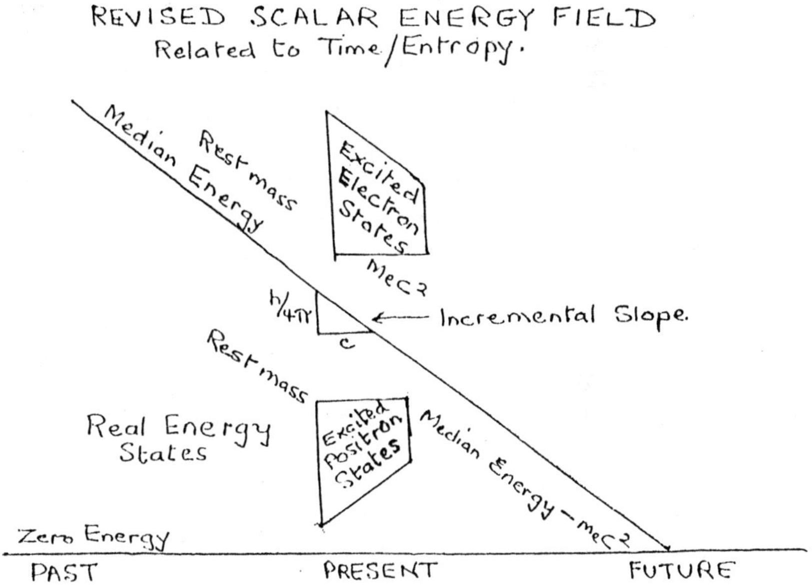

Excited electron and positron states need less energy in time. Bipolar field about median line does not require imaginary negative energy.

the comparison between Dirac's theory, which requires imaginary negative energy states and the alternative, relative to a scalar energy field, where all the energy states are positive and real, when extended from the past, through the present, into the future. The zero point of Dirac's diagram then becomes an average scalar field energy at some point above zero. The scalar energy field, provided by the galactic quasmar, contains all the necessary synchronised attributes of particles. In photoproduction therefore, where half the photon energy raises a particle into existence from the scalar field, the other half depresses the field to form an anti-particle.

If you remember now my conversion table for Newtonian to quantum mechanics you will recall that I wanted to specify position in terms of two phase angles. It is thus only a small step to say that particles above and below the zero energy point are actually in anti-phase. This would place anti-particles at the opposite extreme of the universe, where they may predominate in matter. Anti-particles created in this part of the universe cannot go there instantly, like photons, because they have mass. They are therefore out of place and mutually annihilate when meeting a particle.

The bipolar solution to the quantum field implies that real energy is in a state of oscillation. Real energy can do this if it is a solid rather than a plane wave, something like a wave on the surface of the sea, which has width as well as height. Particles and anti-particles can thus exchange their positions above and below the zero point at a high oscillation frequency but always remaining 180° out of phase. One can thus postulate that every particle has some kind of relationship with its anti-particle at the other side of the universe through the instant transmission of phase in a quantum field. The photon helix can span the universe since it goes through a full 360° change of phase, this is supported by the fact that the photon has a spin of one unit while particles have one half a unit.

It is already accepted that exchange forces causing attraction between particles, sometimes called strong forces, can exist in a vibrating mode without upsetting any conservation principles if the virtual articles that give rise to them pop in and out of existence quicker than they can be caught by the constraints of Heisenberg's uncertainty principle. In quantum theory also, a free particle moving in a straight line cannot be found at certain points along its path, it is as though it either jumps instantly from point to point or flashes in and out of existence. All this is suggestive of the fact that one is dealing with phase relationships in an oscillating scalar energy field.

This scalar field theory has the further advantage that its oscillations occur so rapidly that they do not incur the penalty of a mass equivalent, like the infinite sea of negative energy in Dirac's theory. It is only the slower modulations of particles which are so affected.

In pondering the nature of these presumed solid waves in an energy field that constitute material particles I have to refer back to my thesis in part I, where I suggested that all material particles were held in synchronism by some strong radiation broadcast through the energy field by an object at the centre of each galaxy. Some rule of equipartition of the energy each particle receives must be in operation, to preserve exact similarity of mass and other properties. This, I would suggest is related to the exact width of the time dimension which all material particles occupy in our universe, which is all moving at the same quasi-speed through the energy field, and should all be therefore contracted to the same dimensions by De Broglie's equation. So long as this transition through time is maintained the particle has a unitary existence, it is then only under intense gravitational fields, when time slows down, that the quark constituents of particles may appear.

The actual particle wave group would appear to be a modulation of the fundamental synchronising wave of the enrgy field, and for a solid wave to exist, the fundamental wave must be modulated by two phase elements in quadrature. These phase elements would be of comparable length. The fundamental particle can thus be seen to consist of three vibratonal components, in their various phase states, corresponding perhaps to the three known types of quarks whose various combinations and states make up stable particles. I have hinted that the energy field may have the abiiity to contain resonances and the most feasible explanation of the relationships of particle masses would seem to be that of resonances and harmonic relationships, occurring from the mixing of three wave components.

Now begins the search for the precise wave frequencies in question. If the scalar energy field does contain some fundamental and modulating elements then there would be no probable limit which could be set to their amplitude or frequency. Equal infinities of positive and negative states could exist. The field theory has come upon this problem and for this reason finds difficulty in fitting the field into the finite, but unbounded and curved space of general relativity.

The nature of the vibrational field is such that it stipulates there is 'vacuum energy divergence'. Birrel and Davies

(*Quantum Fields in Curved Space*, Cambridge, 1982),* suggests that, since no particle can carry zero momentum, this leads to a Hamiltonian expression for total energy in which the force restraining vibrations can increase exponentially. Since there is no upper bound to the frequency of vibrations in space it suggests that a vacuum can contain infinite energy density.

In 1975 Professor Hawkinge proposed an ordering operation which demands that, wherever creation and annihilation operators appear, they were to be separated in a left and right hand formulation respectively. This effectively introduces the concept of a bipolar field as described above.†

The further case for resonant modes within the bipolar quantum field would mean that this field was capable of vibration in preferred modes at different points, it would not necessarily consist of a continuous spectrum of frequencies and energies, but instead a succession of line programmes, broadened by modulation with the information related to different physical worlds in separate programmes.

The line programmes on distinct resonant carrier frequencies would have to be separated from each other, rather like transmitted television or radio programmes, in order not to produce mutual interference. This would enable the co-existence of different programmes in the scalar energy field.

At the present I can see no other cause for such a resonance in what is a chaotic background of waves in the energy field, than the particular width of the time dimension of any three-dimensional material universe, determined by De Broglie's equation, as it rushes through the energy field. The length of the wave group of a particle, determined by its speed, would thus have to bear a simple harmonic relationship to the synchronising vibration of the galaxy.

CHAPTER EIGHTEEN

HARMONY WITHIN THE QUANTUM FIELD

Most theorists have a conception of alternative universes, or anti-worlds. The method of reaching them has often been seen as a method of transitting the wormhole beyond the singularity of a black hole in space and emerging from a white hole in another universe. Complex Kruskal diagrams have been produced to track the path a voyager through such dangerous waters would have to follow to emerge safely on the other side. These schemes are however, geometrical constructions which involve systems of four, five and even six dimensions, which are impossible to imagine.

In the quantum field configuration there is no such strain on the imagination. A particular universe, or manifest physical world, would exist around a fundamental tone set by the resonant mode of the field at a given point on its energy gradient. This world system then inhabits this particular band of frequency and amplitude above and below the datum zero energy point, something like an amplitude modulated radio station on the medium wave band.

All lesser states of energy/frequency appear to this particular universe as 'negative energy states', prevented by quantum considerations from manifesting as particles.

Likewise higher energy states would suffer attenuation in the same way through lack of resonance, and only be perceived at much reduced energies. One thus has a scalar field analogue for distance and the spacelike and timelike distinctions of a geometric time cone. In such a system it will readily be appreciated that the visible time cone only extends into the past, which is of higher energy, but attenuated. Scenes from the future are not visible being of lower energy, and negative energy cannot manifest in our universe.

The classical model for a succession of resonant modes in a scalar field would follow a relationship something like that of a musical scale. Using the analogue of a violin string for the helical photon lines existing in free space, then the length and tension of the string sets various harmonic modes which can be preferentially excited by the bow, representing the wave of matter passing through the field. The length of the bow is here determined by De Broglie's equation for the wave group.

A musical scale is determined by octaves over which the frequency doubles. This has a parallel in the atomic table of elements, where progressive increase of the electric charge of the nucleus fall into harmonic groups of properties as successive shells of electrons are formed around the nucleus.

A typical scale of seven notes varies as 1; 9/8; 5/4; 4/3; 3/2; 5/3; 2. The tone interval is 1/8 or 1/16 for a semitone. The violin bow produces a set of integral tones on the violin string. If there were a musical relationship for the family of various elementary particles within a scalar energy field, as the integer relationship of quantum mechanics suggests, then its tonal structure would relate to dimensionless physical constants in nature. The fine structure constant is something of this kind, its value being $\alpha = e^2 / \hbar.c$. This constant is related to the perturbations of quantum electrodynamics and is proportional to powers of 1/137. An integral relationship is compatible with the idea of solid waves.

The precise harmonic mode of a particular universe could be seen to set certain values of its particle sub-fields. The precise graduation of spin in units of $h/4\pi$ defines an incremental difference in the slope of the energy gradient between one instant and another. This is because the distance is determined by c, the speed of light multiplied by the time, which in quantum terms is given by $d\nu$ which is 4π E/h, e^2, the amplitude of the wave.

The decrease of the energy field in time due to entropy means that between two points in the field, one earlier and the other later there is an energy gradient. As the velocity of phase is infinite in the energy field, phase wavelengths only exist in relation to physical space, so that there is no grouping of waves, merely helices of infinite length. Within these helices average wavelengths have increased due to the reduced frequency, so that in the prescribed unitary length of the field there is one less oscillation taking place.

The period in which the minimum energy vibration takes place is $h/4\pi$ .c or 1.581×10^{-26}. Inverted this gives a frequency of 6.325×10^{25} Hz. This suggests itself as a value for the quantum field vibration for elementary particles. For instance, if this is divided by the energy/frequency of a proton, or 2.271×10^{23}Hz, it gives 278.5 (about 2 x 139). A proton wave group would therefore be made up of around this number of vibrations of the quantum field.

It can be established from Hubble's law that after about

2×10^{17} seconds the quantum field vibration would be halved. This corresponds to a Doppler recessional shift of 0.6c, at which point the spectra of distant stars have their frequencies halved. If one divides this time of 2×10^{17} seconds into the field vibration frequency, coincidentally, one gets a value near to c, or 3×10^8 Hz. This implies that the frequency of the field vibration falls by this amount, or c, every second.

While the foregoing seems to indicate that there is a minimum unit of length and time in our universe, the existence of Heisenberg's uncertainty principle written as,

$$d\,x.\,d\,E \sim h/4\pi$$

already foreshadows such a limitation. This does not mean that the scalar energy field itself is in any way limited.

Returning to the analogy of a modulated broadcast programme for the particle wave group, the depth of the modulation of the energy field has some similarity to the quantum mechanical concept of energy wells which contain quarks in a particle and nucleons within the atomic nucleus. The biggest of these energy wells centres around iron in the atomic table, with an atomic weight of 56 and the largest binding energy for its nucleons.

Maybe iron is delving nearly to the zero of the energy field in terms of modulation. The total binding energy involved is 8·75 mev per nucleon, or the equivalent of about half the mass of a proton, or 959 electron masses. The mass loss due to the energy binding quarks in a nucleon must be many times the rest mass of a proton to account for its stability and the depth of its potential well in the energy field.

Quarks are postulated as the building blocks of material particles and are now believed to be inseparable from them and having no independent existence. The properties of quarks include fractional charge of plus or minus one third, and plus or minus two thirds, or zero. Three quarks are thus required for protons and electrons, as a minimum, but the fractional charge need not relate to masses.

I have suggested earlier that in a gravitational field of very high intensity the quark structure of neutrons might break down and result in the re-emergence of charged sheets of quarks in the object described as a quasmar. In such a situation the depth of the energy well generated by the gravity of the qasmar has to fall below that of iron, causing the iron nucleus to break up, and below that of the neutron in turn. I have now put forward the idea that the wave groups of protons/nucleons contains perhaps about 278 wavelets of the synchronising radiation of the energy field. This would lead one to guess that a nucleon might be crushed down to this sort of fraction of its Compton wavelength, by the alignment of its quark constituents. The quark now begins to look like a very thin fellow which somehow weaves a structure of waves to look like a solid particle. This brings one to the new concept of strings which will be the subject of the next chapter.

Before leaving the modulated carrier idea as a basis for the all-pervading energy field it is of interest to note that, while photons emitted by atoms as the result of orbital electron energy changes, and gamma rays emitted by the nucleus itself having various excited states, all follow quantum rules for integer values of energy emission, they are nevertheless subject to small variations due to spin and magnetic moment. One would therefore expect that a scalar energy field would likewise be subject to minor variations as the result of being modulated by a synchronising particle wave. That, for instance, there would be a spread of bandwidth of the carrier commensurate with the depth of modulation.

The degree of spread of bandwith is dependent upon the ratio of the modulation frequency to that of the carrier wave. As a result of this proportionality the upper sideband is of slightly less extent than the lower one. To take a simple example, a broadcast system which operated with a frequency of ten megacycles modulated by a lower frequency of one megacycle would have a spread of sidebands from 9 to 11 mcs. This is a reduction of a ninth and an increase of an eleventh. The difference between these two fractions is 2/99. This has to be measured in units of one tenth of a megacycle.

This does not create difficulties unless one has a minimum unit of length into which the system had to be fitted. They are therefore of importance at the microscopic level of particles, where energy must fit into an integer value. If such proportional differences existed in modulated waves they would have to be accommodated by altering the scale, or by a manifestation of some kind of compensating energy, perhaps in the form of a particle?

The slight difference in energy would mean that the sidebands formed by modulation, consisting of sum and difference waves, would not precisely cancel out. There would be an energy remainder.

The question one can now ask is whether the slight difference in mass one finds between charged and uncharged particles actually reflects the effects of modulation of a scalar energy field. Energy differences exist in all cases, ranging from 1.2 mev in the case of the proton and neutron to 4.59 mev for the pi meson.

The mass of an electron is a convenient unit for the scale in this exercise, being 0.511 mev, this energy is considered to be solely related to the charge the electron has, and therefore all the other mass differences, related solely to charge, should be multiples of it. This is not so for the neutron which has an additional variation in magnetic moment.

On the basis that the proton modulating wave is 1/278th part of the carrier frequency one would expect to find a particle having a mass of 1/77,283 of a proton. Since the mass ratio of the electron/proton is 1/1836 this suggests there are quarks within a proton at around 1/43 of the carrier frequency whose mass is reduced considerably by binding energy.

The derivation of precise harmonic relationships between the quark elements of particles would further reinforce the idea of a modulation in the energy of the scalar field, but this must await further experimental evidence.

CHAPTER NINETEEN

STRINGS AND FIELDS

The attraction of two charges of opposite sign by a static electric or magnetic field is usually depicted by force lines terminating on each charge or pole. These force lines are said to have the tendency to shorten themselves, either along their path or by flowing through interposed conductors or material which offers less reluctance. In any case the present convention does not draw a distinction between + or −, and north or south seeking lines. The diagrams are therefore reversible.

Similarly like charges and poles repel each other. The same convention of single lines do not join in this case, but instead push each other away. Quite clearly the convention is inadequate to indicate the two kinds of behaviour. Nothing is added to the diagram to show why the lines have changed their minds. Clearly the line of force is not something as simple as the convention depicts and something better is needed in order to allow a full understanding of the phenomenon.

Firstly, since the lines are flexible and of varying length this would suggest that they are something like an open helix or spring. The lines have the ability to recognise whether another line it meets has the same or an opposite charge, and act accordingly. That is, the lines join and pull, or don't join and push.

Now if all positive charges emitted lines which joined, then two positive similar charges would not repel each other. If all negative charges emitted lines which repelled, then two like negative charges would repel, but positive and negative charge lines would not know what to do with each other.

So it is necessary to allow that each kind of charge can put out both types of lines. In addition each of the two types of lines put out by positive and negative charges, lines must have the ability to recognise + or − qualities, and be able to decide whether to join or not.

One thus concludes that there are four types of lines of force.

This situation can be simplified if the helical structure of the lines is accepted, together with the idea of a wave impulse passing along the turns of the spring in one direction, in the manner of Fleming's right hand rule.

The lines can distinguish each other by the direction in which the helix curls. A left-handed helix line will join a right-handed one coming in the opposite direction. If a left-handed helix meets another left-handed one they tangle with each other and repel. The joined helices tends to contract if the impulse passes from one to the other, and tend to lengthen if their impulses are opposed.

The wave impulses need not be thought of as emanating from and disappearing into a point charge or pole. It is easier to imagine the impulses passing through the point charge, the types of force line can thus be reduced to only two, each having a direction of impulse into or out of the point charge.

This proposed helix structure for force lines has a direct relevance to string theory, since the universe is now presumed

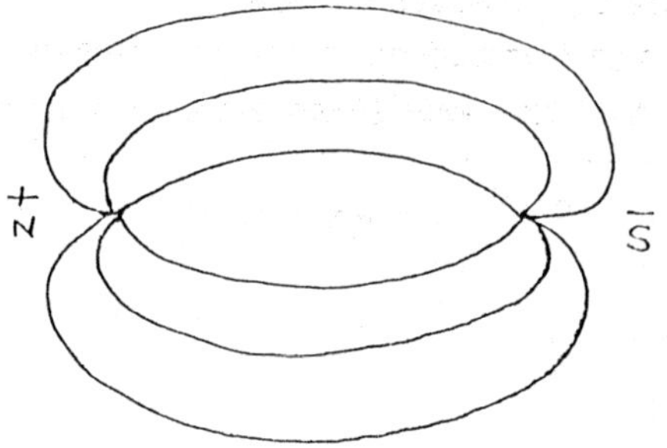
N+
S–

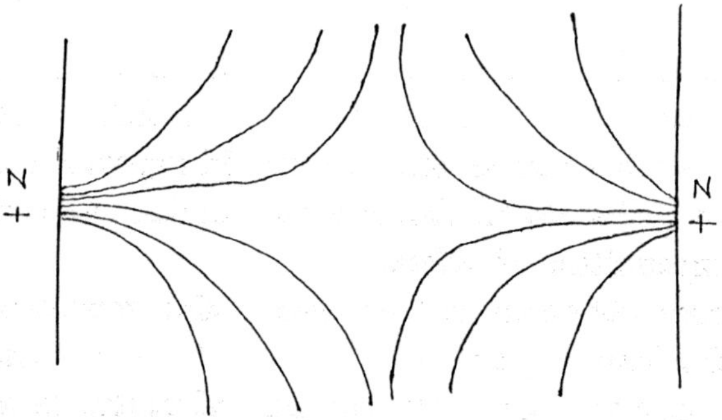
N+
N+

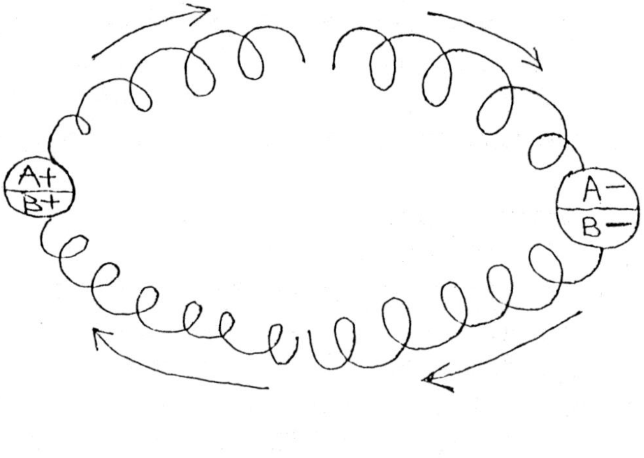
A+
B+
A–
B–

to have originated in thin strings of great density. The helix structure now enables one to understand how they could exhibit complex properties, and it is in accord with Clerk Maxwell's equations. If you remember, these equations provided for a double helix of magnetic and electric force, with the two components mutually opposed at all times.

To recapitulate for a moment, the complex properties which strings exhibit are said to result from the collapsed dimensions which are contained within them. We have now come across a number of ways in which dimensions can collapse:

1. Gravitational collapse at high density and pressure leading to black holes and cosmic strings;
2. Relativistic collapse along the dimension of motion, occurrring as one approaches the speed of light;
3. De Broglie collapse of waves along the dimension of motion, which is linear;
4. Magnetic collapse, or kinking of space, due to the movement of charge at high speed, producing magnetic lines of force;
5. Electrical collapse due to the 'pinch effect' of high current densities, producing electrical lines of force. (Now to be described).

The movement of a double helix of electromagnetic lines through a magnetic field poses a problem. The magnetic component lines up with the field, like a magnet, but the electrical component will experience a kinking force, which would cause the axis of the helix to curl in a circle.* This result puts a restriction upon the movement of radiation through a magnetic field, which would separate it into its two component parts.

This is essentially the mechanism of photoproduction, the creation of a particle and its anti-particle from one photon. It is also the reason why photons cannot be deemed to move along with a three-dimensional material universe but only through a quantum energy field. When the photon is entrained and accelerated by contact with a magnetic field arising from the movement of a charged particle it breaks into two, component particle loops, if it has sufficient energy. If not, photons will just curl and break into two halves of equal energy, or two photons will join together to form one of twice the energy. This phenomena is called 'down compression' or 'up expansion', because when it occurs, in connection with laser emission, the quantum noise background is altered.

The phenomena of photoproduction, down compression, up expansion and laser emission are all quantum field effects, not catered for in the classical theory of radiation. This suggests that radiation is itself wholly a quantum field effect existing in an eternal space with only a transitory contact with our space.

The idea has interesting possibilities in relation to the big bang scenario, which postulates the prior existence of radiation, for one can see how a universe of light can gradually be converted into a mobile and time-conditioned material universe of particles by the simple injection of a few charged particles to seed the process. Their magnetism would produce the knock-on effect of photoproduction from the photon helix background and gradually bring the material cosmos into being. This process may still be going on, as advocates of a steady state theory would have us believe. On the other hand the high-speed particle wave may slow down, and there would be an explosion followed by degeneration once more into eternal, static radiation helices.*

The Behaviour of Current Loops

It is perhaps necessary at this time to add some explanation of the behaviour of current loops, since one has defined above how such loops came into existence through interaction with a magnetic field.

I have suggested that one should imagine lines of force as a flexible helix with an impulse flowing through it. The impulse may be considered as a current, perhaps not of electrons but of some other charged entity. When the helix is bent into a circle this 'current' has a return path around which to flow. This circular current then creates a magnetic force at a point P, for example, situated vertically above the centre of the loop at a distance x. The force is given by:

$$F = 2\pi r_1^2 . I/x^3$$

This force then diminishes as the inverse cube of the distance x, and can be increased as the square of the radius of the loop.

There is a further force created by the current circling the loop. This is a pinch effect, which is a mechanical pressure tending to reduce the cross-section of the loop. This pinch effect increases as the square of the current density. The unit of current, the ampere, is itself defined as that current which can produce a force of attraction of 2×10^{-7} newtons per metre between two parallel wires a metre apart. A current conductor may be considered as made up of a number of parallel filaments which tend to crowd together, producing the pinch effect. The conductor can be ruptured by this effect, as one sees when a fuse wire overheats due to a high current. It can also be the cause of instability in a liquid or gaseous plasma conductor.

The total pinch effect force is given by:

$$F = 2.I^2/\pi r_2^4 .$$

It is an effect which therefore increases very rapidly in small conductors. One can now see that it may be the raison d'être of elementary particles, as a force confining them within certain boundaries. Given that the current carriers, or quarks, remain constant in various species of particles, to give a constant combined level of charge, then the diameter of the string or helix conductor would have to vary to keep the force constant. Energy stored within the loop, derived from the two above equations, would then decrease with the area of the loop but, increase as the square of the current and inversely as the fourth power of the radius of the string forming the loop. The current falls as the area of the conducting string decreases, but increases as the loop shortens. The net effect of all these changes could account for the inverse relationship of particle size to energy/mass.

The increasing curvature of a loop independently exercises a force which reduces the diameter of the string. One can imagine in the primeval fireball, high temperaturs give rise to high velocity collisions which caused compaction of the string loops. At some point the pinch effect takes over and produces a sudden shrinkage of the loop into a stable form which is in balance with the energy field, consisting of a deep energy well related to this pinch effect.

We have so far only considered two frequencies in relation to the functioning of particles – a synchronising vibration from the energy field and a modulation of this which creates a wave group. One is able to construct a flat string, or helix loop, based on this idea, which contains say 278 turns for a nucleon. But the world of particles occupies three dimensions. In my conversion table from Newtonian to quantum physics I had to introduce three elements to define position – two phase angles and a time lapse as the analogue of distance. Clearly we need a third element to define a solid particle.

Let us take the simple analogy of a TV screen. Here two scanning frequencies are necessary, those of field and line, to produce a raster which completely fills the screen in two dimensions. It is not difficult to imagine a spherical screen whose surface could be completely filled up by the same technique, giving the appearance of a solid ball. What is missing, however, is a third component which determines the size of the ball. The two scanning frequencies do not need to differ greatly in frequency in order to fill up the screen by a progression of scanning lines from corner to corner instead of straight across the screen.

I have already introduced the idea of a modulated carrier wave which gives rise to two unequal sidebands, differing slightly in frequency. It is therefore necessary to have two such modulating wave components which can supply the means to fill out the elementary particle in three dimensions. The existence of these two, integral, slightly differing wave components within a particle explains much of the difficulty that exists in drawing up harmonic schemes for relationships of particle mass and other properties. Variations in magnetic moment can also be explained by slight differences in the shape of multiple loops, which would not be perfectly circular, or be exactly aligned.

It will be seen that I have reached a formulation for three components for stable particles by quite a different route from that of the Gell-Mann quark theory, which was based on the analysis of variation of spin, charge and decay schemes of particles.

Let us spare a moment to consider how this formulation of particles relates to their charge. I have already suggested a new description of lines of force of two types, each formed in helices. These lines pass through a point charge. This is logical if the point charge is actually a small loop or loops. One can then picture these lines passing into and out of the particle loops. Since there has to be two types of lines for each species of particle and anti-particle, to embrace the properties of positive and negative charge, there is therefore a requirement for a minimum of two current loops to produce a charge field around a particle. If one equates current loops with quarks, then mesons, the carriers of strong force between particles, are indeed supposed to have two quarks as a minimum requirement for their charge state. If these two loops are considered to be joined, then mesons would look like bent figure eights. The loops would necessarily be of unequal size, since having equal loops would be forbidden by the Pauli exclusion principle. These unequal loops would make the particle rotationally unstable.

Stable particles, consisting of three quarks, would then have three joined loops, again not exactly circular, and of differing cross-sections. The smallest loop would be the largest in mass, while the other two, of nearly the same size, would be of larger cross-section and smaller masses. The division of charge into units of two-thirds and two of

Quark Loops in Particles

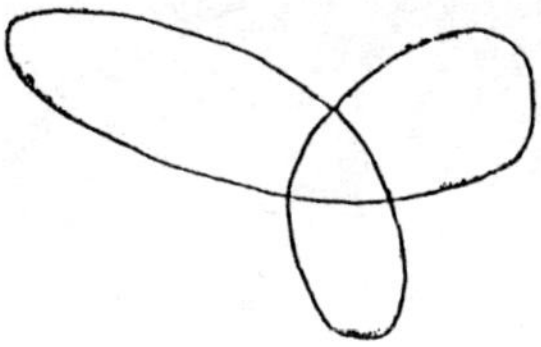

proton/electron

neutral meson

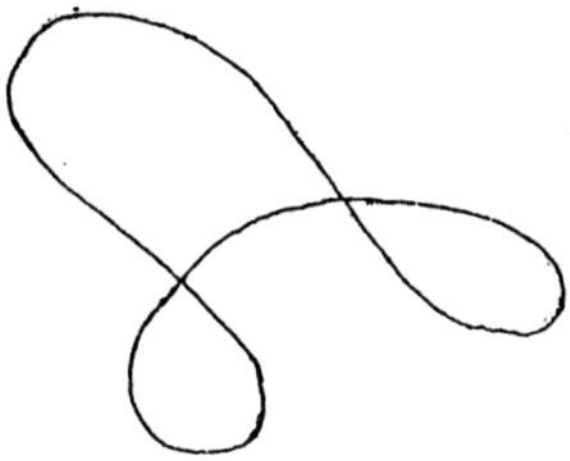

neutron

charged meson

The Nucleon as Three Quark Loops

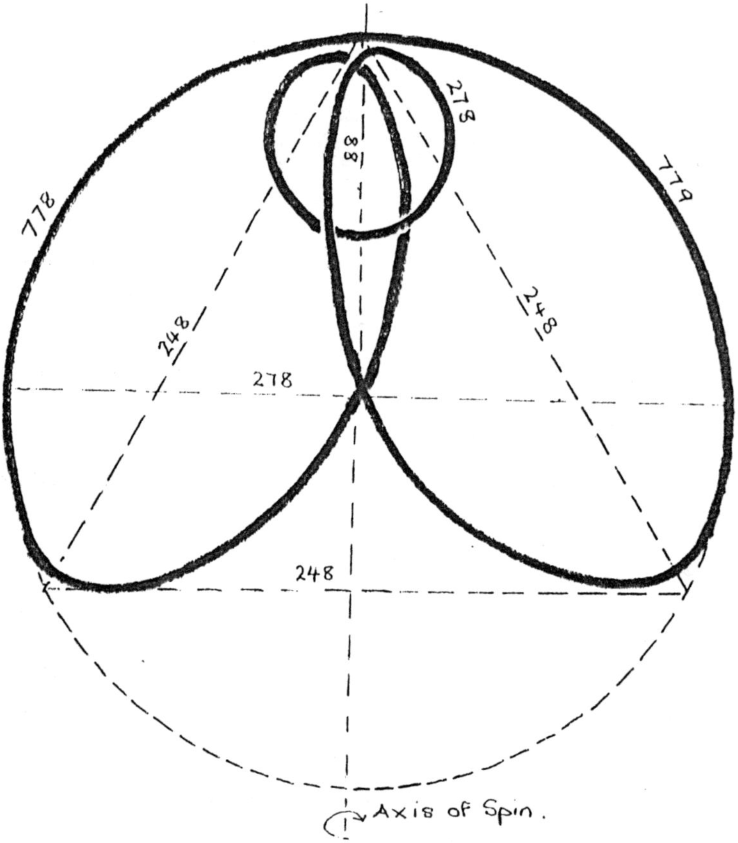

Lengths in units of energy field vibration.
Quark circumferences = 278 + 778 + 779 = 1835.
Cross-section of nucleon = 278.
Binding energy = mass defect = $\frac{278}{88} + 2 \times \frac{278}{248} = 5.4$
in nucleon masses.

one-third each seems reasonable on this basis. Fractional charge can be seen as a consequence of lines passing first through one loop and then another, perhaps around all three to act as a binding force.

In quark theory the 'down' quark is supposed to be 660 electron masses, but a particle smaller than a nucleon has to be more massive. A huge binding energy for the nucleon would account for the small quarks not adding up to a mass greater than the whole nucleon. I have suggested in the diagram that binding energy for the nucleon reduces its overall mass by 5.6 times. The three free quarks would thus be 6.6 times heavier. It would be interesting to speculate that quark cross-sections would be in the ratios 3.4: 1.6: 1.6, so that nearly all the mass defect would be concentrated in one particle. The combined circumferences of the three quark current loops needs to add up to $2.1 \times \pi \times 278 = 1836$ in order to create the harmonic difference frequency of the electron.

The huge mass defect of the nucleon explains why the energy equivalent of the nucleon and its inertial mass does not vary appreciably between ordinary gravitational fields and free space. It needs the pressure and density of a black hole to cancel this amount of binding energy.

The various combinations of current loops goes some way to explaining the varieties of baryons, mesons and leptons. If one has the additional parameters of current loops to work with then these can be equated to quark flavours, which presently consist of charge, isospin, strangeness, charm and colour. In this respect the current loop idea has some flexibility. Isospin, for example, can arise from a slight difference of two modulating frequencies in quadrature. Colour with direction of lines of force, and the energy differences between the force lines due to the differing loop sizes. Strangeness and charm can be accounted for simply by adding further current loops for SU(3) and SU(4) groups of particles (see table in appendix). The many flavours of gluons, up to eight, can be deduced from the coupling of the loops and the way in which they are bent, which alters the degree of their mutual coupling.

The current loop explanation of the slight mass difference between the two states of the nucleon – proton and neutron, which is currently accounted for by transfers of charge and magnetic moment, can be further pictured as a rearrangement of the cross-points of their three loops, and thereby the direction of the force produced in them. Two opposed loops thus give rise to a closed, or neutral field, in which all force lines thread internally. For the anti-particles the current flows in the opposite direction around the whole figure. In mesons the single crossing point goes under or over to produce a neutral particle, while in the charged meson the loops do not cross.

Is this just another loopy theory?

CHAPTER TWENTY

THE STRONG, ELECTRICAL AND WEAK FORCES

The force of gravitation has a very long range. Electrical forces usually operate over shorter distances because of the equalisation of charge which occurs in large masses.

Electrical repulsive forces between like charges increase in proportion to the inverse square of the distance up to the surface of the charge sphere of an elementary particle, at which point they fall off linearly in proportion to the distance. At this point nuclear attraction takes over, as a manifestation of the strong force, until finally a repulsive core is reached where the yukawa potential takes the form,

$$E\rho \,.\, (r) = -E_o \,.r_o \,.e^{-r/r_o}/r$$

although there is some difficulty in expressing this qualitatively as a repulsion.

How can such behaviour be explained by the current loop model of particles, where such particles are postulated to have a pseudo-velocity relative to a quantum vibrational field?

Firstly one must assume that the spacelike dimension of eternity contains magnetic lines, as discussed later.

Secondly one must assume that charge consists of particles drawing together magnetic lines of force, or the static helices of the quantum field, as it transits this field at high velocity. These lines pass through the particle loop and spread out again. The loops tend to be saturated, because of of their high velocity, and only a slight attenuation of the energy field occurs at long distances. This explains the slight attraction of gravity as that of an energy well produced by this attenuation, resulting from mutual shadowing of large masses against the energy field. As two particles approach closer to each other they both tend to draw lines together from the same cross-section of the quantum field – this constitutes an attractive or repulsive force according to the respective signs of their charges.

At some point, when the lines spread out, after passing through these particle loops, they will be caught up by the loops of the neighbouring particle. The double absorptions and emissions of these lines will interfer with the electric force while at the same time increasing the depth of the energy well between the particles. This would mark the boundary of the charge sphere, where the electric force ceases to operate, and a strong nuclear force of attraction is substituted.

A further point will be reached where the particles approach very closely and no more lines can be threaded from one to the other, since such lines would then have to bend at a sharp angle. A closed, bipolar field will be produced at this point which will tend to cancel out the energy well between the particles and keep them apart at the repulsive core.

The sharp angle at which lines of force are bent at this point of close approach suggests that it might be the formative process for mesons, as virtual particles, and the carriers of the strong force. Mesons decay very rapidly

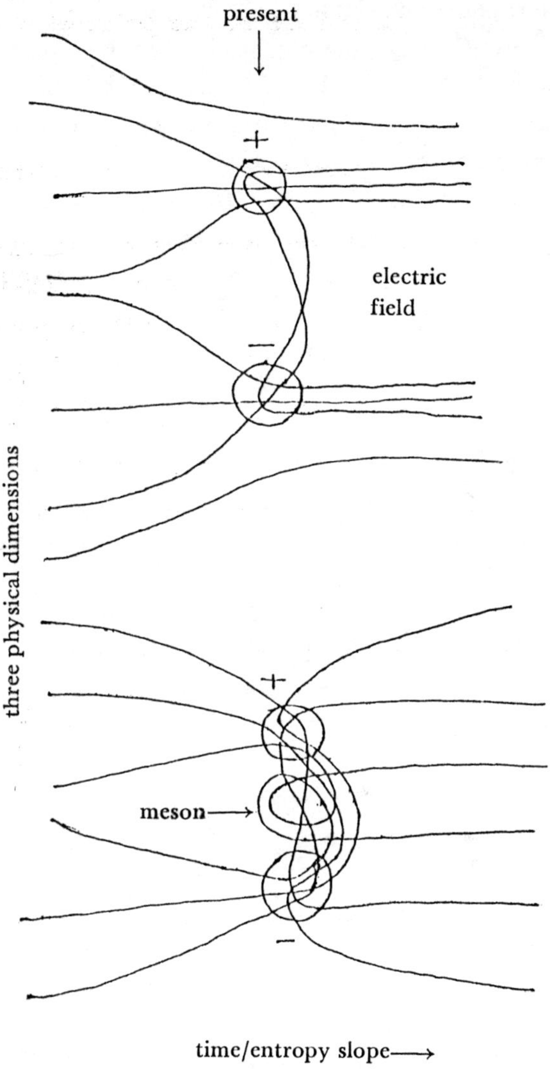
present
+
electric
field
−
three physical dimensions
+
meson
−
time/entropy slope

when separated from particles which could be due to a rapid unbending of their force lines when released from inter-particle constraints.

We now have a possible model for the combination of electric and strong nuclear forces. Would it be possible to extend the same model to include the weak forces of gravitation and radioactive decay?

Considering first the weak forces which are associated with radioactive decay and the stability of nuclei, these seem to apply to elements heavier than bismuth, with an atomic weight of 209 and an atomic number 83, and also to lighter isotopes which depart from a mean proportion of charged and neutral nucleons, or which have high rates of spin.

One can assume perhaps, that in these unstable nuclei, the depth of the potential energy well for the ground state of the nucleus is insufficient to stop nucleons or smaller particles from occasionally jumping out. It is customary for the particles in the nucleus to be considered as a fermi gas. The fermi model of the nucleus gives a reasonably accurate estimate of the depth of the nuclear potential well. This is measured by:

$$E\rho = 3^{2/3} . \pi^{4/3} . \hbar^2 . \rho^{2/3} /2M$$

Where ρ is the nuclear density and M the nuclear mass.

It is clear from this equation that any decrease in h in relation to time reduces the depth of the potential well and eventually brings about a state of instability. Natural radioactive decay can thus be seen as an ongoing process, not merely dependent upon statistical probability, or half-lives, but gradually involving more of the heavier and less stable elements as time goes by. This concept allows a more understandable explanation of how the heavier elements came into being in the past, when a much deeper potential well would have allowed them greater chance of being formed as stable entities.

This, however, is not a comfortable theory to live with, for it implies that, with the passage of time, further heavier elements will lurch into instability and start decaying with perhaps disastrous consequences. It is not a difficult task to work out the most likely candidates for such incipient breakdown, and to take some measures to isolate them in good time. Bismuth, already mentioned is top of the list with only one stable isotope out of six.

It is of passing interest that throughout the history of the earth the end of each geological age has been marked by 'faunal extinctions'. The last such episode occurred about 63 million years ago when the dinosaurs died out over a period of ten thousand years. The latest evidence suggests that this 'faunal extinction' was not due to the impact of an asteroid or meteor, but rather to increased volcanism and earthquakes, which released gases and dust into the atmosphere and caused a dramatic drop in the earth's temperature, together with acid rain and the pollution of the oceans. What could cause heightened geological activity at the end of each 100 million years? The fact that evolutionary leaps also occur at the same time, with the emergence of new species, goes hand in hand with the idea that there might also be heightened radioactivity during such periods.*

The cause of the earth's sub-surface warming at different periods could be due to new heavy elements becoming radioactive. This is not a serious problem unless one of those elements happens to be fissionable, or able to split into two nearly equal halves. Chance concentrations of fissile material might then produce a planetary disaster like the one that befell the planet which once occupied a position where the asteroid belt now is between Mars and Jupiter.

This is not to suggest that the present phenomena of global warming might in some way be related to the theory I have now put forward. Increased tectonic activity and volcanoes would be the most like symptom.

In radioactive decay a number of isotopes make isomeric transitions, which consist simply of the emission of a gamma ray, so that there is only a slight alteration of mass and the internal quantum state of the nucleus. Examples of this kind of change occur in Na 24; A1 26; C1 34; Sc 46; Co 60; Zn 69; Ge 75; Se 77, 81, 83 and so on. The isomeric transition is said to be a reversion from an excited state of the nucleus to a ground state, which can take seconds, minutes or hours, a very long time on the atomic scale.

Excited states of nuclei can be induced by high energy collisions with other nuclei, but the absorption of energy into an excited nuclear orbit is not like an elastic collision with an immediate rebound. Some very subtle mechanism retains the bundle of energy for a while and then lets it go.

The period of decay for isomerism is measured in half-lives, like any other form of radioactive decay, during which time half of the excited nuclei return to the ground state. This represents a process governed by a chance expectation. A chance of what?

The excited, or higher energy orbits of both electrons and nucleons are generally eccentric, while those of the ground states are more nearly circular. Is there therefore some mechanism which dislikes eccentricity and requires circularity? Such a mechanism might of course be high-speed pseudo motion through a regularly structured energy field. This field accepts certain shapes and sizes more easily than others, circularity in particular. This is the same as railway lines accept rolling stock of the same gauge with little friction. This would be a new twist to existing ideas about gauge theory!

In current particle physics radioactive decay is moderated by a W particle with a mass of about 200 times that of a proton.

This mass is dictated by the distance over which the particle has to operate, or about 10^{-17} m. The reader will recall that I have theorised there might be 278 energy field vibrations within the Compton wavelength of a proton, so what we are considering here may be an effect related to a fundamental string width within our material world. This string width is supposed to be increasing with time.

End of the world scenarios seem to occupy the minds both of religious eschatologists and science fiction writers. Isaac Azimov, in his book *Robots and Empire*, offers a world ending very similar to the one I have hinted at above. He envisaged a species of death ray called 'the intensifier', which operated by increasing the flow of W particles, causing atomic reactors to explode and an increase of natural radioactivity in the earth's crust such that the planet became uninhabitable. This was a deliberate move to make humans colonise other worlds. Reality may be different in that an intensifier principle is built into the system, perhaps to make us leave Eden in due course.

CHAPTER TWENTY-ONE

GRAVITATION, FORCE OR FANCY MATHEMATICS

This manifestation of a weak force is the next nettle that must be grasped by the theorist. One can start with Newton's enunciation of a law of gravitational force, the mutual attraction of all lumps of matter by a force varying inversely as the square of the distance between them and measured as a constant fraction of their mass.

This requires something very unscientific called 'action at a distance', with nothing to create that action in the intervening space. Newton himself did not wish his theory to be linked with such a concept. So long as there was a medium in which electromagnetic waves could be propagated, a luminiferous aether, then gravity waves might also be generated in it, and the lurking anomaly of 'action at a distance' could be shovelled under the carpet. In the pure physics of the nineteenth century one had this perfect trinity:

Matter ⇄ Electromagnetic Waves ⇄ Luminiferous Aether

But then came the experiments of Michelson & Morley in the 1880s, which have already been described, which failed to discover any drag from the aether medium upon the earth.

It was into this theoretical vacuum that Einstein's relativity theory was sucked. He first proposed an invariant speed of light and then suggested an exterior, geometrical framework for space and time, in which the force of gravity was replaced by curvature, due to the presence of mass. This all rested on the principle of equivalence which can be stated as follows:

The Principle of Equivalence

This states that the effects of gravitational force and the accelerations resulting from it were indistinguishable within an enclosed space for a short period of time from accelerations of an external frame of reference.

Take the case of a lift which is falling freely down its shaft. A frame of reference accelerated downwards with the lift would appear to have abolished the force of gravitation within it. Objects in the lift would have no weight. Similarly a rocket in free space, removed from gravitational fields, if boosted to give the correct acceleration, would appear to have a gravitational field acting in the opposite direction to the boost.

Einstein therefore argued that gravity could be regarded as an apparent force due to the kind of reference frame one has chosen in space, and one could have a multiplicity of these reference frames. This view is based ultimately on the *equivalence* of inertial mass, which resists movement, and gravitational mass, which creates an acceleration. This equivalence has been tested to an accuracy of one part in a million million, yet why it should be so remains a mystery.

The effect of the earth's gravitational field produces an acceleration of 9.8 metres per second in all masses, and involves a build-up of energy within them proportional to their mass times the square of their velocity. This gravitational mass is the same as the inertial mass one would have to overcome in pushing any mass across a frictionless, level surface. This is an enigma because the gravitational mass has been balanced out by that surface.

The curvature of space invoked by Einstein also affected electromagnetic waves, in order to give proof of its existence. Magnetism was in addition able to produce an alteration of space, so the above trinity was restored in a different form:

Matter ⇄ Spacetime ⇄ Electromagnetic Waves

As a theorist you have already guessed that I would like to use Occam's razor here and remove the ghostly intermediary 'spacetime', managing with the two outer observables which we know and can measure. Other theorists have not been enamoured of this configuration and propose instead a particle solution:

Matter ⇄ Particle Exchange Forces ⇄ Electromagnetic Waves

This gets rid of the idea of 'action at a distance' completely. Unfortunately we have no detectable candidate for the 'graviton' or exchange particle.

If the graviton exists it would be involved in gravity waves and these are presumed to travel at the speed of light and would have low frequencies making them difficult to detect. The graviton would, however, have to operate over large distances and in view of the high streaming velocities which have been calculated for neighbouring galaxies, in the order of 1,000 km/sec, relative to the cosmic background, this puts some strain on a particle theory. If centres of gravitation are in motion the orbits of planets and stars should be disturbed, the force would be a certain time arriving and would be out of line with the observed position of the star. The sun, for example, would move through a quarter of its diameter relative to space while its gravity travelled to us. This would alter the earth's orbit. Either we have a separate frame of reference in the solar system, as Einstein says, or there is some other teleological principle at work which effectively predicts the sun's position.

Can one now reinstate the first formulation in a slightly different way:

Matter ⇄ Scalar Energy Field ⇄ Electromagnetic Waves

I have proposed that matter has an invariant velocity relative to the energy field while electromagnetic waves have a phase state within it. This means, as we all know, that one cannot change the rate at which time passes in a material sense without attaining a velocity near to that of light, such that inertial mass is also increased. At slower speeds inertial mass maintains its equivalence, because one is diverting matter which is already travelling at a constant pseudo-velocity.

How would a scalar energy field manage to generate a universal form of attraction between all lumps of matter? Clearly with so much energy resident in this field it only requires the syphoning off of a tiny fraction, such that energy wells are created between particles. Such energy wells would be extremely shallow, the difference between the electrical and gravitational attraction between an electron and a proton, for example is 10^{39} times.

Large numbers like these have a fascination, so let us

compare this one with the time the universe has been in existence, or 3×10^{17} seconds and the period of oscillation of the Compton wave of a proton, which gives 6.8×10^{40}. This suggests that gravity is linked to waves having a very long period, and that maybe the scalar energy field contains such waves of very high intensity.

You will remember that I have already postulated that the cosmos is turbulent. This is a characteristic of the hyperconic structure which would not be present in the hyperspheric form.

The questions which now have to be answered is whether there is sufficient long wave radiation in the universe to account for gravitational force, and what is the mechanism by which current loops of particles can extract energy from them.

Electromagnetic waves with a frequency as low as 1 Hertz can now be detected by a process of beating two waves of higher frequency together with a difference of just 1 Hertz. The results of experiments at Arecibo Observatory in Puerto Rico, Los Canos and the Island of Mona, show that waves of ultra low frequency exist at large amplitudes, far larger than can be accounted for by the Stefan Boltzman law governing the energy radiated by a black body. This law suggests that energy radiated at extremely long wavelengths should be vanishingly small.*

Since radio emissions from space have also proved stronger than that predicted by the black body radiation law, so other mechanisms for creating them have had to be advanced. The foremost of them is synchrotron emission, caused by the intense magnetic fields of stars and galaxies, which accelerate streams of electrons. Magnetic storms in the atmospheres of planets are also known to be the cause of radio emissions.

Longer waves may be produced by the solar wind, a stream of electrons which emanates from the sun and stretches as far as the earth's atmosphere. This wind is modulated by the sunspot cycle and by the rotation of the sun, moon and earth.

The evidence for extremely long cycles of magnetic fluctuation is contained in the earth's rocks. Magnetic signatures frozen into volcanic rocks as they solidified through the ages show that the earth's magnetic field reverses about every million years. The magnetic field of Uranus is now believed to be at the halfway point of flipping over.

The Arecibo investigators discovered sporadic waves with a signal to noise ratio of 20 which occurred in bursts of 40 seconds. They concluded that this was due to instability in the thermal movement of charged particles in the earth's ionosphere or alternatively the presence of gravity waves. This noise spectrum is particularly interesting as it rises from 5 Hz and goes off the scale as it approaches zero Hertz. Large unknown mechanisms like periodic plasma heating, are suggested to account for these new observations.

I have already suggested that quasmar oscillators in close neighbouring galaxies could beat together to produce radio frequencies. It is not stretching imagination too far to imagine that repeater oscillators reinforcing the quantum field within a galaxy sometimes beat together to create low frequency waves.

A new law for background radiation intensity at low frequencies may be needed. It is possible that in this relatively unexplored region of the electromagnetic spectrum very high intensities may exist, capable of exerting some mechanical force.

As already stated, electromagnetic fields consist of two elements. While the electric component can be radiated from an aerial length of about one quarter wavelength, a magnetic wave can be propagated from a much smaller current loop. Whereas the electric field obeys the inverse square law of attenuation with distance, the magnetic wave is attenuated as the cube of distance. In contrast to the electric field, which radiates in straight lines from the transmitter, the magnetic field is made up of closed elliptical lines. These factors combine in a way which makes the magnetic field component more potent at low frequency, rapid attenuation being compensated by increased efficiency of radiation, when produced by current loops instead of straight conductors.

ULF waves are difficult to detect because they spread out so far in space. A wave of 0.8 Hertz extends out to the moon, a wave of 7 Hertz encircles the earth. Even longer waves may be present in great intensity, but so little of their energy is measurable at any moment that they escape notice. In this sense the wave with a very long period of months or years has a teleological aspect, it reaches from the past into the future. Waves which extend from the past into the future effectively carve out a path in space for large astronomical bodies. They would, in effect, be following a line of least resistance, Fermat's principle, and like something which Einstein used to define tenors in his geometrical system.

Magnetic fields are in a different category to electric fields because they are the same in a moving system as a system at rest. Magnetic lines derive from the relativistic contraction of groups of moving, charged particles. The motion can be rotary or in a straight line, and force results when the lines of one system cut those of another. Strangely, however, lines of force from a rotating or travelling magnet cannot be thought of as travelling with it, this kind of relative motion of magnetic fields does not exist.

The candidature of magnetic fields for an existence throughout a time continuum would depend upon their presence being detected all through space. Alfven has noted the possibility of their being a background magnetic field of a millionth to a hundred thousandth of a gauss. There is evidence for this in the polarisation of light from distant stars.* Cosmic background noise as stated above is thought to be caused by relativistic electrons which are accelerated through largescale magnetic fields.

To be able to consider gravitational force as a field phenomena existing between two masses, one must take the ultimate case of two separate particles being impacted by ULF magnetic waves and somehow managing to abstract energy from them. The scenario is easier to construct if these particles are given a pseudo-velocity through a scalar energy field. This would result in the cutting of lines of force if they were present in the field.

To understand why there can only be a one-way transfer of energy from the scalar field to the particle one must take the analogy of the Compton effect which occurs when an electron collides with a photon. The energy of the photon is always reduced according to the relationship:

$$1/\nu_1 - 1/\nu = h/m_e .c \times (1 - \cos\theta)$$

This is not a quantum effect. The energy change is independent of the wave energy and only related to the angle of incidence of the electron and the photon. It is arguable that it could be very low indeed, and if the angle of incidence was very small, energy might still be abstracted from the wave in accordance with the above formula. It is reasonable to assume that the force lines of magnetic waves of very low frequencies would be aligned along the axis of pseudo-velocity of the particle, as the angle decreases with frequency.

The experience with heat pumps shows that energy can be extracted from a piece of ground which is already cold by making it still colder. By this kind of analogy we can see that there is no wave in the universe of so low a frequency that energy could not be taken from it to make its frequency lower, and perhaps this is the key to the whole process of entropy and the expansion of the universe as we observe it, since expanding wavelengths require a larger universe to contain them.

Two particles moving at a pseudo-velocity through an energy field and collecting energy from very long waves would find that their zones of collection would inevitably overlap. This overlapping is what would create an energy well between them, resulting in a force of attraction.

Gravitation is a negatively signed force and is opposite to angular momentum which is outwards and positive. In this theory, as gravitation involves the absorption of radiation so angular momentum does the opposite and creates radiation, producing a force of repulsion, to cancel the attraction of gravity. This, then is the complete correlation of gravity and electromagnetism which eluded Einstein. This theory presumes that gravity and angular momentum will always coexist in a mutually self-cancelling mode like electric charge.

Surely, one might argue, gravity is a very exact force, related by a constant to mass, unvaringly accurate to many places of decimals. Magnetic lines of force, on the other hand, would vary wildly from place to place, so would the permeability of space to such lines. But these variations are at low crossing speeds of lines and low currents. Above a certain limit everything smooths out. This is called saturation, no more magnetic lines can be got into the lump of matter. So, providing the particle or mass is moving fast enough its magnetism stays constant.

The force generated by a charged particle is given by:

> $F\rho = -e.v.B = (e.v.\mu_o . I)/ 2\pi a$ where
> a is a constant for each particle; e is a constant; μ_o is a constant; B is a constant. When saturation is reached the current I will be a constant. If the velocity v is also a constant, too, then the force remains constant.

Take the exmple for a shunt-fed electric motor. Providing there is some slight residual magnetism in the core it will deliver electric current when it is rotated. The current it delivers will continue to rise up to a certain number of revolutions per second, when among other things, the core becomes saturated. Then, no matter how much faster it is rotated, no more current comes out.

Next, how does one explain an inverse square law resulting from magnetic lines which obey an inverse cube for attenuation? The answer may be given in various articles on what is called 'mock gravitation' as an explanation for the mechanism by which galaxies were formed in the early stages of the evolution of the universe.*

The theory is that absorbing particles in an isotropic radiation field can experience a force due to their mutual shadowing. If the pre-galactic universe contained enough suitable radiation, and absorbing particles, an instability would be produced because these particles would create dark zones where the radiation was excluded. Radiation pressure on the outside of the zone would then have acted to further concentrate the mass of dust particles.

Radiation pressure can have a significant effect on the structure of the inter-stellar medium, and can dominate gravitational attraction, as when the tails of comets are pushed away from the sun by the pressure of its radiation.

Spitzer was the first to point out that particles could experience an inverse law of attraction due to their mutual shadowing in a bath of radiation. Using his theory, later researchers have found that many observed facts of galaxy formation fall into place, and no fatal flaws have emerged.

A mock gravitation theory which involves the extraction from magnetic waves of low frequency by a Compton style impact may seem a far cry from the present tidy expression for gravitational force:

$$F = G.m_1 .m_2 /d^2 \text{ newtons}$$

Where G is 6.67×10^{-11}, a constant known with considerable accuracy, and is the same for all types of substances, so it must ultimately be based on the uniformity of constituent particles, nucleons and electrons.

The amount of energy absorbed per particle from an ULF magnetic field per Compton impact would be vanishingly small, but of course, this is a criteria for the delicate shading of gravity. The very minuteness of its quantum of action has been the great problem. The magnetic mock gravitation idea requires that the force is built up from innumerable tiny impacts.

If one measures the amount of gravitational attraction between two particles a unit distance apart and then multiplies by 1/G or 1.5×10^{10} the result is their actual mass. The phenomena of mass may thus itself be derived from something of the order of this number of Compton-style collisions in an energy field in unit time.

There is of course no precise way of determining what the background magnetic field would be, unless of course one used as a reference the cosmic background radiation of 2.7° K. In terms of mass this background is equivalent to 4.15×10^{-40} kg. One could then say, for example, that a proton, with a mass of 1.672×10^{-27} kg is raised from this background a factor of 4×10^{12} times. This is greater than 1/G by a factor of 267. This perhaps indicates that the scalar energy field increases the mass effect by this amount. This, of course, is only speculation, but it is near to the figure of 278 I had arrived at as the difference between the Compton wave of the proton and the scalar energy field vibration.

It will be of great interest to see whether any future studies throw light on the possibility of a magnetic connection for gravity. The consequences of it being proved are quite wide-ranging. In the first place, a proper understanding of the nature of gravitational force should quickly lead to devices which can nullify its effects, just as surely as Einstein's equation of energy with mass lead to the atomic bomb. Secondly our ideas about the infinite increase of gravity in a black hole will be changed, since this is no longer dependent on geometric distortion but has to rely on what energy is available in the scalar field.

When the ultimate collapse of a neutron star comes

about the energy well can only dip down so far before it passes outside the present moment of our universe. Figuratively one can see that, when particles are grouped too close together they begin to exert such a drag on the energy field that they get left behind in time. A wormhole then opens in our spacetime which spreads back up the energy gradient until it reaches the previous turn of the hyperconic spiral and funnels into an earlier galaxy. This process has already been suggested. There may well be 278 turns to this hyperconic spiral, each spanning fifty million years. Between the turns of this spiral there is only eternity. But never fear, the ultimate end is order and the resurrection of a new star system.

CHAPTER TWENTY-TWO

CHAOS OR QUANTUM COHERENCE

Chaos theory seems to be in vogue at the present time because in many ways it reflects what we observe around us in the forces of nature. Chaos occurs in a small way at first, as a tiny aberration, somewhere at the boundary between the field of quantum mechanics and largescale phenomena, but then it grows and grows, leading to bigger irregularities in otherwise stable systems, which is the signature of chaos.

Chaos represents a state of persistent instability engendered from small beginnings.

In the region of largescale phenomena one is accustomed to think that the future is uniquely determined by the past. There must be causal links which explain everything that happens. The Newtonian laws of motion ought to be able to predict a course which bodies in motion will follow. But now it seems, order and determinism of this sort is not universal.

The original mathematical equations which outlined the possibilities of chaos were invented by the French mathematician Laplace in the eighteenth century. Although a strict determinist he was interested in predicting things like the fall of dice, as gambling was popular then as now. He even applied his ideas concerning probability to human behaviour in random situations.

In the nineteenth century the two systems – determinism and probability existed together and seemed incompatible. It was not until quantum theory came along that probability could be used as an exact method to predict transitions in large numbers of particles. Yet even with larger systems some element of chance operates where more than one force is acting upon an object.

The usual method of illustrating chaotic effects is to have a magnetic pendulum bob acted upon by two magnets positioned below it. When the bob moves to a point midway between the two magnets its future motion becomes extremely sensitive to small changes in its present position and velocity, so its motion becomes chaotic. If the error in measuring the present position and velocity is as small as a tenth of one per cent, this error can be amplified after only one swing by up to ten times, and this error increases by the same factor at each swing. After three swings there is 100% error. To get any accuracy of prediction at all over nine swings therefore, one would have to be able to measure position and velocity down to the size of atoms, and here one enters the realm of the uncertainty principle in quantum theory.

It is easy to see that Newtonian style predictions become more difficult, even on an astronomical scale, where large bodies have been in orbit over immense periods of time, subject to perturbations by the attraction of neighbouring bodies.

Chaos theory is now used in the study of aurorae in the earth's atmosphere, caused by charged particles being trapped by the earth's magnetism. Weather forecasters use the Laplace equations for prediction. Biologists can estimate things like the spread of diseases by chaos theory and engineers now have to take it into account in the design of large structures.

At the end of the nineteenth century another Frenchman, Poincare developed a new kind of mathematics called topology which has been developed into a powerful method for finding connections between varying quantities. This was developed by later French workers and has lead to the beautiful modern fractal pictures generated by powerful computers in America. These show how very complicated patterns can be evolved from quite simple variable equations.

It has been shown that systems which involve feedback of part of the output of a system to its input can also engender chaos, since such a system may amplify small variations to as great an extent as the desired signal or movement. A system with several stages of amplification therefore resembles the pendulum after a number of swings. This is a question of great importance since much of human activity depends upon feedback of sensory information to the musculature, and it would seem we achieve greater accuracy by this feedback, rather than descending into chaotic behaviour.

The question that now concerns us is whether chaos can in the end enter the region of scalar energy fields and completely overthrow the rule of law in the cosmos at large.

I would like to propose therefore that there is a counterbalancing arrangement which I have called 'quantum coherence' which holds chaos at bay. It has some relationship to Rupert Sheldrake's idea of a morphological field, which prescribes that similar complex systems may interact, and presumably would do so through the agency of the scalar energy field.

The effect of quantum coherence can be assumed to depend upon the conjunction of several factors, for exmaple:

1. Similarity of structure and function;
2. The existence of alternative, metastable states within a structure, capable of storing information;
3. The duration of the alternative states. The longer this is the better the chance of coherence;
4. The idea that the quantum of action may be related to time, making every instant distinct from the previous one.

Most of the above criteria can be deduced from the phenomena of coherent radiation. When certain atoms absorb radiation their electrons move into excited orbits. Normally an electron can return to its ground state in one step by the emission of a photon, but in some cases it has to do this in two steps by the emission of two photons of different energies. This takes longer than the energy absorption process, so it is possible to pump radiation energy into the atom where it will be stored for a short while.

When a number of atoms are in the excited state they can be triggered to drop into a lower state together. This simultaneous action results in the emission of a single pulse of coherent, monochromatic radiation, referred to as laser action. It can be seen how laser action depends on the fulfilment of conditions 1, 2, and 3 above.

Going back to first principles in quantum mechanics, the Heisenberg uncertainty principle prescribes that, on the atomic scale, the position, Q, and momentum, P, of a particle do not necessarily commute. That is, $Q.P - P.Q = i.h$. This leads to the construction of the time independent Schroedinger equation:

$$i.h\, d/dt\, |\psi(t)> = H\, |\psi(t)>.$$

Where H is the Hamiltonian operator for a system evolving in time. This determines the way in which the 'state vector' $|\psi(t)>$ for the system has an increasingly probabilistic factor as one reduces things down to the size of elementary particles.

Morphological coherence appears to defy these rules by a species of 'quantum mechanical' relationships for large structures. This means that chance expectation can form the basis of behavioural rules, in the same way as for elementary particles.

This extension of the rules of chance expectation would seem to depend in a linear fashion upon the number of information-storing elements, or logic states, as in 2 above. Likewise the smallness of the volume, r^3, within which these elements are contained. Smaller size will clearly promote more rapid action and greater coherence. The duration of the refractory effect, t will also increase the potential for coherence in a linear manner, as in 3. Lastly, coherence is enhanced by simultaneity, which involves rule 4. The shorter the period in which the action occurs, the sharper will be the value of h. So an expression for quantum coherence can be written on this basis as:

$$i.\hbar.t_1\, .n/r^3 = H/\psi t_2 \rho > \ldots.$$

Where ρ is expressed in units of pascals, or pressure

This may seem an odd result until one considers that electrons in excited orbits do exert a form of mechanical pressure upon neighbouring atoms, and that the triggering of laser action can ripple through a mass by a variation of contact pressure.

When one considers the vast amount of genetic information contained within a living cell, the existence of quantum coherence between cells with similar DNA in a living organism is possibly explained, such that mutual co-operation and dependence is assured. The behaviour of insects in a hive follows this pattern.

Human brains also could enter the realm of coherence, having up to 10^{10} nerve cells, some with as many as 10,000 synapses and a refractory period for the storage of information of many years, all within a volume of 0.00150 cubic metres, which should satisfy the necessary criteria.

If the reaction of large numbers of neurons are linked in such a way as to reflect the conditioning of the brain, then this confers a measure of predictability, and coherence with persons having similar conditioning can be expected. Hence crowds of people tend to produce stereotyped behaviour, as at football grounds. The violent aspect of such behaviour can be due to a kind of psychic pressure, enhanced by close proximity to other reactive structures. Rupert Sheldrake also emphasises the 'critical mass' of common thought which is necessary before action results, as in the hundredth monkey effect.

Coherence factors have been calculated for a number of comparative systems as follows:

For an atomic system with an electron triplet decay time of 10^{-5} seconds, a radius of 5.3×10^{-11} m and n = 1, $\psi\rho = 6.67 \times 10^{-9}$

For a living cell with a diameter of 20 microns, a refractory period of 10^{3} s, possessing say 5 million gene codons,

$$= 5 \times 10^{-10}$$

For the human brain with a diameter of 20 cm, having 10 cells, in the neo-cortex, each with 10,000 synapses, capable of retaining information for many years, the factor works out at

$$\psi\rho = 6.67 \times 10^{-9}$$

The fact that the brain has a coherence close to that of the near certainty of atomic structures, and better than that of single cells, may be another reason for cell co-operation in larger organisms. Mere size of brain is not enough, it is the complexity of the structure that counts.

In such cases there is a simultaneity factor, or $, which can be inferred from the almost instant transmission of thought patterns, which does not rely upon spoken or any other obvious form of communication. In the case of atoms it may well be a pressure effect which produces the unerring repeatability of laser action.

The factor $ is in some ways similar to the Q factor in a resonant oscillating circuit, which acts as a multiplier of voltages. To produce a dimensionless factor of this kind one needs to relate two successive values of h at the start and the conclusion of a process. The first value of h is slightly larger. The expression could therefore take the following form:

$$\$ = 1/(1 - h_t^2 \; / \; h_o^2 \;)^{1/2}$$

This will give a large value of $ when the two instants close together. We now have two expressions which confer a different meaning on time dependence. Although the simultaneity factor is wave related it does not develop like a wave function. It nevertheless indicates how probability can be conditioned by coincidence. This coincidence has to occur at the origin of each wave, and so is related to the phase state of the system.

It could be argued that quantum mechanics deals with observables in particle and atomic states but says nothing about the coincidence or otherwise of phase between separate atomic structures. Such coincidence is measured by relative angular velocities of electrons in orbit, which can be known exactly, while the possibility that several atomic systems might get in step at a certain point of time is unknown.

Insights into scientific theories sometimes come to us in strange forms. This last one occurred while I was watching a TV advertisement for Californian raisins, in which a line of 'raisin men' performed a dance to the tune 'I heard it on the grapevine', while they whirled their arms in a series of complicated movements, almost too fast for the eye to follow. But, at a certain beat of the music, they all stopped and pointed with their arms. This dance exercised a strange compulsion on my imagination until I connected it with the idea of coherence in the phase relationships between atomic systems. The suggestion that coherence existed as a form of pressure had hovered in my mind, but had remained unresolved until a way had been indicated in which such pressure might act.

Electrons orbiting an atomic nucleus in bound energy states have a condition of phase relative to one another as regards their angular velocity. This phase state is disturbed as the result of an electron being boosted into a higher and more eccentric energy orbit when it receives a photon. The electron moves under the influence of a central force varying inversely as the square of distance, so that the radius vector sweeps equal areas in equal times, while maintaining a constant total energy. Nevertheless the electron does seem to slow down and point at furthermost extension of its elliptic orbit. Conceivably the phase relationship which would cause many orbiting electrons to point their orbits in the same directiton could give rise to a pressure effect in that direction. The pressure effected by coherence is then not uniformly distributed over the surface of the mass, but is preferentially orientated in one direction, and critically dependent on phase angles.

Since one is dealing with an inverse square law there is no quantisation of phase states in related systems, which can therefore be observed to have an exact coincidence in time.

It is therefore apparent that the higher the frequency of a cohering system, the greater the accuracy with which it can be directed. This possibility is therefore of interest in connection with the design of a scalar wave transmitter, as described in the appendix.

While atomic systems have little difficulty in matching their mutual reactions to the heartbeat of a galactic scalar field vibration in order to realise coherence, it would appear that our brains would have greater difficulty in doing this. Precise control of breathing and automatic reactions would be necessary in order to project scalar field energy in any desired direction. Nevertheless the potential exists for us to achieve strange powers in this field if we practise the technique.

We may be at the point of discovering a new field of mechanics. In addition to classical mechanics for large structures, quantum mechanics for very small ones and chaos theory at the interface of the two, there may now be a new relationship dependent upon complexity, which one might call holistic mechanics. Here the probability of action is dependent upon both coherence and simultaneity, and takes the form of a pressure directed linearly through a scalar energy field which resonates instantly with similar structures at a distance.

CHAPTER TWENTY-THREE

ETERNAL INTERFERENCE

The human brain contains in all about 60 billion cells packed close together with an intricacy of white matter, or myelin wiring between various cells. Each neuron in the brain will build up a potential of a few millivolts and fire at around ten times a second in the absence of stimulus. There is thus about twenty watts of electrical energy being expended inside our skulls, mostly in the form of heat, for as long as we are alive.

With all this radiation being emitted at close quarters there seems ample opportunity for the most of it to be diffracted within the cerebral interstices. Patterns of interference would then ensure that about 30% of this energy would end up being cancelled by interference.

I have early supposed that radiant energy cancelled in this way does not disappear, but turns up elsewhere, due to the factor of altering the zero of the energy field. If these ideas about the displacement of the zero energy point turn out to be correct, then it is easy to see why our thoughts could leak out and be picked up by others. These patterns of neural activity could turn up in odd places instantaneously. It would be no wonder that we were enjoined not to think dangerous thoughts!

The remaining 70% of the radiant energy in the brain,

being converted into photon line helices, might be considered our legacy for eternity, since the myriads of such lines remain frozen into the energy field once the physical brain has passed by on its merry way to dissolution.

If something can be left behind in the past, then equally a bit can trickle into that part of the spacelike dimension which lies in the future. Then one is left with the awe-inspiring consideration that our neural pattern may be imposed on some unsuspecting foetus at a future time.

Life itself may arise from another kind of interference with the scalar energy field. The double helical form of DNA molecules may have more than a coincidental relationship to radiation helices. The activity which DNA exhibits in connection with the replication process, splitting and transferring its messages to RNA, is not adequately explained on a chemical basis. Some more subtle energy may be drawn from the scalar energy field by the helix, whose construction might act as a receptor. Molecular vibration also produces effects like catalysis which seem to derive energy from a sub-atomic source.

Lest anyone might suppose we have stumbled upon a non-polluting energy source capable of supplying all our needs, let us remember that nature would not have given us stomachs and lungs if we did not require chemical energy to supply what was necessary for the body in respect of muscular activity. If nature could have tapped a scalar energy field for large physical requirements, would it not have done so?

APPENDIX ONE

DESIGN OF SCALAR WAVE OSCILLATORS AND RECEIVERS

This apparatus would be designed to shift the zero energy point of a scalar field in such a way as to transfer information rather than actually produce any local catastrophe.

To do this one needs to produce fields oscillating in anti-phase, which ostensibly cancel one another out and therefore transfer their effect to some other point. Three anti-phase systems are required, each positioned so that their polar radiation patterns are mutually at right angles and therefore spreading equally in all directions.

Take the case of a transformer with one input winding and two similar output windings. If the outputs are connected in series and in anti-phase the output voltage will be zero. The self-cancelling effect of the two windings occurs because the middle phase point of zero moves up and down with the alternation of the voltage. The driving electromotive force therefore has no zero reference to push against.

Providing no current is drawn from the secondaries, the current input will be the same in both cases. An electromagnetic field is radiated in the first case. In the case of the anti-phase connection one cannot cancel energy, so where has it been transferred? If radiation cannot occur then transposition to an appropriate state of the scalar field is a possibility. Since the scalar field has point correspondence with the physical universe then it seems logical to infer that energy would not be broadcast but moved en bloc to a particular plane. Using two transformers in quadrature reduces the displacement to a line. With three transformers mutually disposed at right angles the displacement is reduced to a point.

In the case of these transformers the zero shift is not held constant, so the point-to-point transmission would move around and give strange effects in the locality. These have been noted by experimenters who have dabbled in scalar wave transmitters.

A practical scalar field transmitter therefore has to have sophisticated tuning and phase shifting devices which can offset a zero shift by a predetermined amount, while providing a fixed direction for the transmission by means of phase shifts.

Any kind of suitable oscillators can be used but the anti-phase aerial system should be omni-directional, this is because the scalar field comes from and emanates to all directions in space. For example, if one is using current loops at low frequency three of these should be arranged mutually at right angles. Phase shifting two of the elements will then provide directivity for the third loop.

The energy shift of the zero of the scalar field would be measured in Hertz. The amount of energy required depends upon the total rest mass energy of the matter enclosed within the volume of space where the field is to be modulated, multiplied by the frequency shift, and divided by the field vibrational frequency.

The wattage of alternating current that a core of laminated iron can handle without heating is usually measured by its cross-section. However, since current is being passed through coils disposed on three axes, this wattage must be reduced by at least three times. The best material for low-frequency work is probably iron dust.

The cross-section of the core required for a given wattage is given by:

$$A = (W)^{-1/2} / 0.865, \text{ where } A = \text{area in sq. cm}$$
i.e. a core of 1 sq cm would handle 0.75 watts

The number of turns for the three separate windings can be calculated from the formula:

$$E = (4.44 \times F \times H \times N \times A) / 100{,}000{,}000$$

Where E is the voltage, F is the frequency, H is the number of lines per sq cm. As there are three separate windings at right angles to each other the usual number of lines, which would be 7,750 should be reduced to 2,500 for each coil. This means winding on more turns.

The mass of a core of dust iron of 1 cu cm is 5 gm. The total rest mass energy, $m.c^2$, equals 4.58×10^{14} joules. To convert this into watts one must multiply by the field factor of 278/G (this is the rate at which the mass is built up and maintained by travelling through the scalar energy field). This is then divided by the field vibration frequency, and gives a figure of 30 watts per metre/Hertz, as a measure of the energy required to produce a given displacement divided by the frequency of the oscillator.

The experimental core at full power should thus produce a displacement of between 2 and 3 cm multiplied by the oscillator frequency, which can be used for the calibration of the effect.

Diagram of Transmitter Element for Scalar Wave Emission

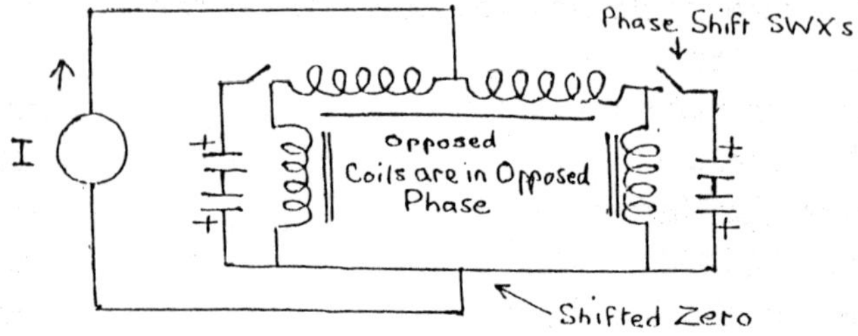

Coil Windings

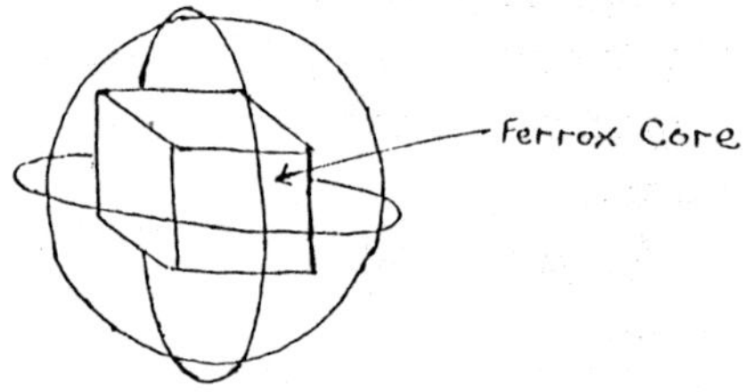

Receiver Current Loops

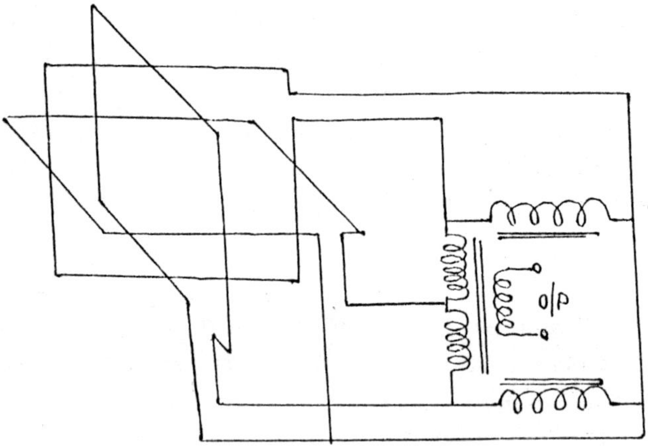

It will be apparent that efficiency increases the smaller the transmitter coils can be made and the higher the oscillator frequency. The same would apply to receivers.

History

The earliest experimenter with scalar wave generators was Nicholas Tesla in the 1920s. He used large crystals and very high voltages to produce the oscillations. The piezo-electric properties of crystal were discovered by W.G. Cady and published in 1922. The frequencies were in the high audio range and his coils generated a powerful magnetic field.

Quartz crystals are commonly used for generating electromagnetic oscillations where a high degree of stability is required. The resonating properties of a crystal are determined by its dimensions, the way the exciting voltage is applied to the crystal planes, and six elastic constants. The activity of the crystal is greatest when electrical stress across two of the crystal's faces produces a large mechanical stress. It is the mechanical vibration that keeps the oscillation going, and it is probably this effect on the environment that enhanced the scalar field effects.

The difference in the propagation characteristics of quantum fields has also been elaborated by Henley and Thirring in connection with a study into how the pressure waves of ultrasound are transmitted when the separation of air molecules is greater than the sound wavelength.* There are then no waves of compression and rarefaction, but instead a condition called 'zero sound'. The sound no longer radiates in concentric circles, but instead there are 'spin waves' according to the formula:

$$(s - \cos\theta)\,\nu(\theta,\varphi,\alpha) =$$
$$\cos\theta\ 1/2\ T\ r_a \int F(\mathfrak{x})\,\nu(\theta',\varphi',\alpha')\ d\Omega/4\pi$$

A variation of the centre of oscillation relative to point $p = 0$.

Taking the simplest case: $F(\mathfrak{x}) = \Phi(\mathfrak{x})$, $\Phi = \Phi_o$; when the Fermi surface is stretched out in the direction of motion.

High-frequency ultrasound does not therefore propagate in air and ultrasonic cleaning baths need to have the transmitting crystal attached to the bath. Nevertheless, standing near the bath, one's teeth can be set on edge at certain angles due to a species of linear transmission of the ultrasound as described in the above formulas. This is a good analogy for linear energy transfer in a scalar energy field.

Crystals are commonly cut into plates such that voltages can be applied across them at angles of 35.5° to the 'Y' plane. This is called an 'AT' cut and the resonant frequency is given by the equation:

Frequency (Hertz) = 1675 thickness of plate (metres)

To produce radio frequencies it can be seen that a plate of only about a millimetre is needed and a potential of a few volts. The impedance of quartz is around 7,000 Ohms per centimetre. So, in ultrasonic baths, where crystals of around 2 cm are used, several thousand volts are necessary to produce significant mechanical vibration.

In order to get effective oscillation from the very large crystals Tesla used in his experiments, very high voltages, possibly several million, would have been needed, and the resulting vibrations would have created something like a minor earthquake.

There is of course a 'Q' factor involved in the resonance of a crystal. This multiplies the peak voltage of the energising circuit by a factor of around 2,000. This would be a further reason why strange, sidereal effects resulted from experiments in the low frequency region.

Recent research has centred on the production of low energy, solid state oscillators, which operate at brain wave frequencies around 15 Hertz. It is claimed that these devices can partially entrain beta waves and keep someone alert when carried on the person. Users remark that the square wave pulses appear to register inside their body.

APPENDIX TWO

WHY SHOULD ONE OPT FOR A HYPERCONIC UNIVERSE?

If there was an initial big bang which sent matter in all directions at or near to the velocity of light, then two observers A and B, who evolved a long while after that event, would not be able to see each other until their time cones overlapped. If their initial trajectories diverged by more than a few degrees this would not occur until millions of years later. On the diagram, time is plotted vertically and the other dimensions of space are horizontal. The further the observers look out into space the more the time cone curves down into the past.

If the two observers diverged at more than 60° they would never see each other! A section of the universe with this solid angle of divergence is therefore totally separate for all practical purposes in a flat universe, and might as well be an independent entity, a hypercone.

You might think that anyone at the edge of the hypercone could see beyond it, but this is because the picture only represents a three-dimensional view. In four dimensions each observer is at the tip of his own expanding, three-dimensional time cone.

In a curved space you might arrive back at the site of the big bang eventually, but in curved space there would not be any single bright spot in the cosmic background radiation.

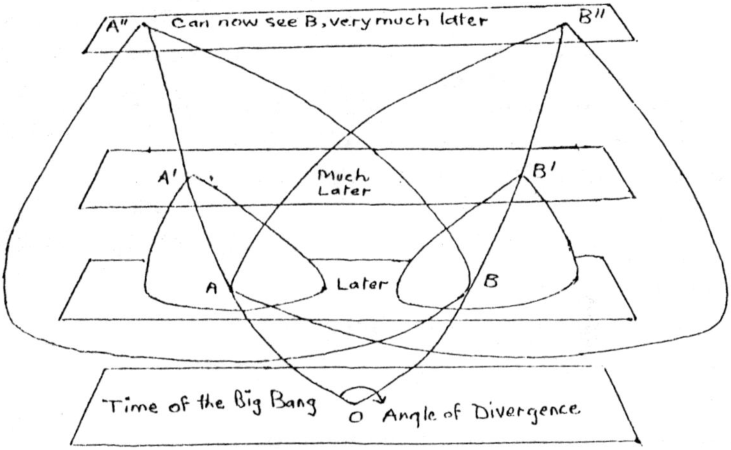

See also Appendix Eight, page 203.

APPENDIX THREE

THE TRAMPOLINE EFFECT

The laws of least action and least time were mentioned earlier in the context of unexplained laws of nature. I have now introduced the idea that the force of gravity is required to be predictable if orbiting bodies are to retain their correct angular velocity around highly mobile primaries. In furtherance of this idea it was necessary to propose that gravity was caused by the interaction of matter with ULF magnetic waves which exist throughout spacetime. The wavelengths would be very long, projecting many millions of miles through a conjunction with the present moment into the future.

The physical co-ordinates of a wave state which occurred one second ago are separated from us by a distance of 3×10^8 metres. Likewise the co-ordinates of an electromagnetic event which is to happen one second from now are at a similar distance from us. These electromagnetic events appear to approach us from all directions at once from a future state, and having focussed momentarily in the present, move apart equally broadcast in all directions into the past. This is because we can have no dimensional correlate for time except along our personal time cones.

The aerial transmitting the electric component of an electromagnetic wave needs to be of a length comparable to the wave to obtain reasonable efficiency. This rules out the possibility of interaction with waves many millions of miles long. A magnetic wave, however, can be radiated or received by a magnetic loop, which concentrates the flux. This kind of action is suggested for the model of the elementary particle, and likewise opens up the possibility that matter can operate to exercise a species of teleology over future largescale events, so that for example, the course which the earth will follow during the next few seconds can be accurately predicted. Its inertia and angular momentum allows one to do this. In fact, an ephemeris can be drawn up which predicts astronomical events many centuries ahead.

This argument has its parallel in the world of particles, albeit relating to shorter instants of time. Yukawa's meson theory allows virtual particles to exist within the fraction of an instant allowed by the uncertainty principle and provides the basis for the action of strong forces. This shows that time can be taken by the forelock, so to speak, and that the inverse side of the coin can apply to large masses. What this would imply is that largescale mechanical action has a predictable outcome, curtailing the recent ideas of chaos theory.

The image which I would like to draw for the reader's benefit is that of a very large trampoline, whose linkage extends either side of the present moment into the future and the past. The present moment therefore stretches in a line across the spring framework.

If a weight is placed upon the trampoline along the line of the present this will be distributed through the spring linkage sideways through linkages disposed in the future and the past. As time travels across the trampoline the line of the present advances carrying the mass with it, at its pseudo-velocity through the energy field. The dent in the trampoline thus moves with the weight and projects a valley of least action along which any reaction will proceed.

In the geometric frameworks proposed by general relativity, where the presence of matter produces dents in spacetime, a primary like the sun necessarily is required to have its own system of co-ordinates in order that its energy well can remain static within them. The curvature of a four-dimensional spacetime can then produce its equivalent of gravitation by curvature tensors of the fourth rank, $G\mu\nu = 0$, which prescribes a form of motion in the energy well which incurs least action.

In this present thesis, the existence of very high streaming velocities of galaxies relative to the cosmic radiation background, and possibly also relative to the big bang, cannot be ignored, even though they are not obvious by current tests. Neither can general relativity account for the occurrence of least action in chemical reactions and all kinds of processes which cannot conceivably have separate inertial frameworks.

The trampoline principle is intended to exemplify a dynamic and all-embracing process related to matter, derived from the pseudo-velocity of all matter through a scalar energy field. The spring linkage is the equivalent of magnetic lines of force, which can account for the apparent attraction of gravity, and valleys of least action in all material processes. It possesses an entelechy not realisable in general relativity, but compatible with Maxwell's laws, which amongst other things, has the ability to generate a rotational impulse exactly opposed to the magnetic/gravitational attraction. It is axiomatic that both electric and magnetic forces should call into existence an opposing force by induction.

While Fermat's principle of least action and de Maupertois' principle of least time have been given mathematical expression by Lagrange equations and assumed in Feyman diagrams, it is an unsatisfactory state of affairs not to have them explained by any theory. How does a ray of light on reflection know the path that will involve the least time and energy? The same applies to all other reactions which follow this rule in nature. The trampoline principle shows how local energy can be evenly distributed to produce valleys which guide the reaction, just as someone jumping on a trampoline will produce depressions which guide their footsteps. No part of the linkage bears a disproportionate load, each carries its exact share, no more no less, and each tries to regain its former undisturbed state. The degree of accuracy of load distribution depends only on the fineness of the mesh, which relates to the uncertainty principle, and within these limits one can prophesy the future course of events.

APPENDIX FOUR

ENTROPY AND THE HEAT DEATH

Entropy is a term which crops up in thermodynamics. Heat is a measure of the energy present in matter and entropy is the part of that energy which cannot be used.

The second law of thermodynamics states that the entropy of a closed system cannot decrease – heat energy must therefore pass from a useful state into a useless form in order to deliver work. This is certainly true of a steam engine with its Carnot cycle, where one starts with high-pressure steam in a boiler and ends up with warm water. One can only extract work energy from the engine in one direction.

In another sense, entropy is a momentum towards disorder and chaos. In the case of the internal combustion engine one must first use energy to refine crude oil into a metastable petroleum, such that it can be easily exploded under pressure and produce power in the engine. The resultant exhaust gases are a disordered and useless mixture of steam, oxides and nitrides.

If entropy increases in a general way throughout the universe, and this applies equally to the nuclear engines in the stars, then everything is running down and becoming disordered, more and more useless, and the end will be the heat death prophesied by Arthur Eddington, where everything throughout the universe is slightly warm.

This second law is the charter for all doomwatchers and scientific pessimists. It would discourage all but the most masochistic from a further study of physics.

The idea of entropy however, glosses over some little anomalies which are generally related to a change of state. For example, some substances, on cooling, can, like water, pass into a crystalline state, like ice, which is greater in volume by 10% than water. Ancient quarrymen used to pour water between cracks in the rocks so that freezing would split them. At certain specific temperatures therefore, change of state due to cooling can be used to produce energy. This also occurs by using certain refrigerant gases in a heat pump.

In 1972, J. Beckstein examined the conditions relating to the occurrence of entropy in a black hole in space. This black hole is produced by the gravitational collapse of a star, as mentioned elsewhere in this book. The idea was that there could be an analogy between physics (gravity) and thermodynamics (temperature), in that, the event horizon surrounding a black hole, which is the boundary beyond which light from the central singularity cannot pass, would expand with the accretion of matter by the black hole.

One could say that matter has fallen into a more useless and disorganised state within the event horizon, if it cannot get out again. Beckstein concluded that the entropy of a black hole was proportional to the size of its event horizon. This lead to the further conclusion that the black hole should have a temperature, and that this temperature was inversely related to mass, as the event horizon got larger the temperature fell.

I have mentioned elsewhere how, in 1972, Prof. S.W. Hawkinge described the quantum process by which black holes could radiate photons. The uncertainty principle then permits the generation of particle pairs by high gravitational force fields in a very small interval, the particle/anti-particle pair could subsequently become separated by magnetic fields. Such a process would eventually lead to the collapse of the black hole.

Another effect which has not been considered however, is a possible change of state due to the low temperature. We have seen how, during the process of stellar evolution, hydrogen is converted into helium, generating energy in the interior of stars. Heavier elements, up to iron on the atomic table are also created by the fusion process at high temperatures and pressures. Then we see the process being partly reversed when the star turns into a white dwarf, because matter becomes degenerate and the electron orbits just collapse in on the nucleus. Then there is a transition to a neutron star under still greater pressure.

Is this the end of the story when the black hole forms? Could there not be a further intermediate state when matter reverts to a quark plasma under further increase of gravitational pressure? More importantly, do these final stages of neutron star and quark plasma represent further disorder and entropy? Or is something akin to the formation of a crystal likely to occur, which would constitute a more ordered state. Both the degenerative and plasmic metamorphoses are designed to occupy less space, so, in other words, can order be induced by close packing?

There is, of course, the final temporal change in relativity theory, which occurs when gravitational pressure is so great that the progression of time is slowed down to a standstill. General relativity does not regard such a condition as an asymptote, and presumes matter can still exist when time is at a standstill. In fact, physicists suggest one can go beyond this point and have time running in reverse. This means that a black hole is inverted into a white hole, emitting matter somewhere else in the universe.

On the other hand, quantum theory suggests that the four forces of nature tend to merge under extreme conditions. Extending this idea to include temperature would necessarily produce asymptotic conditions in an energy well. This is because the temperature cannot fall below absolute zero. As a black hole expands and cools, it freezes to within one ten millionth of a degree of absolute zero. The quark plasma is thus frozen hard and solid, and when time ceases to progress it must revert into radiation in accordance with the scalar field theory I have advanced.

At each stage of conversion from a star to a white dwarf, to a neutron star and to a quasmar, there is an explosive situation. The difference with the black hole is that radiation cannot escape from the hole, it must therefore reconvert into particle pairs, neutrons and quark plasma. What one has within the event horizon therefore is an *oscillatory state*, and the vibration of the quantum field is the only remaining method by which the black hole can shed its energy accretion.

One does see analogous oscillation at the onset of degeneracy in cepheid variable stars, but when the neutron star phase is reached, the oscillation has been stabilised by high-speed rotation. Further collapse should only accentuate a process already begun, and therefore the quark plasma star should be postulated as an entity rotating at very high velocity, which enables it to maintain the stability of its oscillation between states by the generation of a very intense magnetic field.

In the light of previous arguments about the hypercone structure of the universe within spacetime, and the progression of matter through a scalar energy field, the progress of a quasmar will be constrained by the close packing of matter, thus leading to the relativistic slowing of time. There will also be a tendency for linear motion to be converted into angular momentum, due to the structure of the hypercone.

The penultimate state of matter in the quasmar is then a natural sequence of events from the neutron star, only rotating more rapidly and oscillating more energetically to dispose of the accreted energy of infalling matter. This is a well-balanced system and the occasions when the ultimate collapse into a black hole occurs should be comparatively rare. When that happens, the hole will become massive enough to swallow a galaxy.

The penultimate state of matter in this theory is therefore not disorganised, useless energy. Entropy is held at bay in most cases and the quasmar becomes a potent source of synchronising energy, capable of maintaining and synchronising the existence of myriads of stars in the galactic systems and their environs. This is a much more hopeful doctrine for astrophysics and is one which parallels many other renewable cycles in nature.

APPENDIX FIVE

THE COSMIC FAMILY OF FORCES*

Strong Nuclear Force
Considered to be the force binding atomic nuclei; about 100 times as strong as electric force; short range, acts at 10^{-16} in on quarks; carried by mesons. See page 143.

Electric Force
Otherwise defined as the electromagnetic interaction; it binds atomic systems of particles; movement of electric charge creates electric current and magnetic fields by altering the spacing of the charges relativistically; strong at atomic distances, around 10^{-11}; weak over long distances, since matter as a whole is electrically neutral; acts on all charged particles. Treatment of lines of force, page 143.

Weak Forces
Responsible for radioactive decay; acts on all particles, including quarks; strength about 10^{-11} of electric force; very short range 10^{-17}m; carried by W+, W–, and W° particles. Supposed to act by charged or neutral currents. See page 146.

Gravity
Keeps the planets, stars and galaxies in their places; acts

on everything, matter, radiation, energy; weak in atoms, only 10^{-39} of electric force, strong in dense matter, collapsed stars; infinite in range; obeys an inverse square law and so is not subject to quantum laws. See page 150.

Colour Force

Otherwise known as the chromodynamic interaction which binds quarks in elementary particles; it has strong nuclear force as a by-product; acts on quarks; strong at the range of 10^{-15} m, the size of a proton. See page 141.

Gauge Theory

This says there is a pattern to which particles must be matched, like matching successive rolls of wallpaper to each other on a wall. Galactic systems of particles must conform to one pattern. This, then, is a constraining force, based on mathematic theory. Is it due to the scalar energy field? See page 149.

Angular Momentum

Said to be a constant for the whole universe, present at the time of creation. It can be used up in one body by friction, like the effect of the moon's tides on the earth, but this lost momentum is transferred to the moon. See page 20.

Inertia

A tendency of matter, once set in motion, to continue moving in a straight line unless influenced by forces. According to Mach's principle it is due to the interaction between all matter in the universe. See page 1.

Professor Eric Laithwaite of the Imperial College of Science & Technology, England, has invented a machine for nulling the effect of gravitation as illustrated opposite.

This machine consists of a central, upright rod which is spun by means of an electric motor at its base. Towards the top of the rod are two smaller opposed rods which are free to move up and down laterally through about 90°. At each end is an axial, air-driven, brass gyroscope. When these gyroscopes are spun with a blast of compressed air their movement causes the hinged bearing rods to rise and revolve around the main spindle due to precession effects. This revolution is aided by the electric motor at the bottom.

A curved rail fixed to the motor housing catches the end of the gyroscopes' bearing shaft as they rotate, and forces them down through 90° before releasing them. The reaction to this downward force of the curved rail provides a lifting thrust in a cycle which is repeated as the main shaft rotates.

Professor Laithwaite insists that he has no quarrel with Newtonian laws of motion, but he does suggest that they need modifying.

Newton, it seemed, rather overlooked the problem of gyroscopes, and their tendency to do unexpected things. The laws of motion should therefore be modified to account for gyroscopic precessions.

The secret of the Professor's machine lies in the fact that no energy is required to move the gyroscopic arms upwards after they have been released by the descending, curved rail – gyroscopes do this naturally as they precess. It does not contravene any conservation law, because the upthrust energy is derived from external current to the motor and the air blast for the gyroscopes.

Post-Newtonian mechanics would also suggest that this device is a powerful source of gravity waves – since the weight of the gyroscopes is augmented by their angular momentum.

Note for experimenters: The heavy construction illustrated is a measure of the considerable vibrational stresses likely to be generated. Damping of natural resonant vibrational frequencies is advisable.

APPENDIX SIX

MAGNETIC MONOPOLES

Grand unified theories which seek to link the electromagnetic, weak and strong forces have theorised the existence of a gauge field, which has been here variously compared with matching patterns of wallpaper design and rolling stock running on railway lines of the same gauge.

This gauge field is supposed to interact with a Higgs field, which causes it momentarily to lose its symmetry and break down into the several observed forces of nature. In this particular theory the theoretical breakdown is the result of the pseudo-velocity of matter passing through a scalar energy field.

In 1974, Gerard t'Hooft and Alexander Polyskov showed that the simplest prototypes for such a theory allowed twisted configurations of the Higgs field which acted as the source of an ordinary magnetic field – in short they were magnetic monopoles. Physicists have now derived exact solutions for the fields of several t'Hooft/Polyskov monopoles which fit various gauge theories. These solutions are very complicated as the field of one monopole is distorted in the presence of another. However, with the aid of one of the latest developments in computer technology, the transputer, it is now possible to process a programme simultaneously with many different variables in parallel, and so speed up the solution of the problem immensely. This work was carried out at the University of Southampton by Tony Hay, Chris Jesshope and Denis Nicole.

It is now therefore possible to depict graphically how the energy density field varies as two monopoles approach one another. Images of the energy distribution have been modelled in three dimensions with the aid of the Winsom programme devised by the IBM Scientific Centre at Winchester.

This modelling shows that, as the two monopoles approach one another their fields evolve in a peculiar way. It is not symmetrically distributed about the axis joining the two monopoles – it appears instead to cross over like a distorted figure 8. Also, when the monopoles are nearly coincident, there is symmetry about a curved axis which at either end is perpendicular to the axis of approach of the monopoles. In effect, both diagrams appear to be 'wind blown', or distorted by a pseudo-velocity through a resisting medium.

Both models, according to field theory, relate to a transition from weak to strong force, which I have attempted to describe in this book. (Refs. *New Scientist*, 17.9.87, p. 49 and 16.10.86, p. 53.)

APPENDIX SEVEN

COSMIC COMMUNICATION

I mentioned earlier that the hyperconic explanation of an expanding universe does allow communication between its separate parts sufficient to ensure a uniform type of development. I have also suggested that the lapse of time should not be thought of as movement through an actual geometric dimension, but as progress along an energy scale (though for convenience one must draw time as a dimension for purposes of illustration).

The hyperconic model presumes a generalised expansion in three geometric dimensions at or less than the velocity of light. Depending upon the vertex angle of the hypercone, signals sent between observers experience a progressively larger time lapse as the universe expands, because this expansion carries them away from each other in time and distance from the origin of the big bang, where everything was absolutely synchronised.

Let us take for example two observers A & B, equidistant from the origin O to which they are joined by dotted lines which signify duration. When they try to communicate by light signals a further time lapse occurs which carries them both forward in duration through time in the line A,A' & B,B'. The light signal which goes between them is also carried along with the time lapse from A to B'. But a geometrical expansion of the universe also increases the distance between A & B to A" & B" in one direction, A"' & B"' in a second direction and A"" & B"" in a third. The light signal sent from A to B therefore must travel in a spiral path in all three dimensions to reach B"". This will apply to all signals which do not originate from the big bang itself. Spacetime therefore has an inherent hyperconic structure which is independent of the presence of mass.

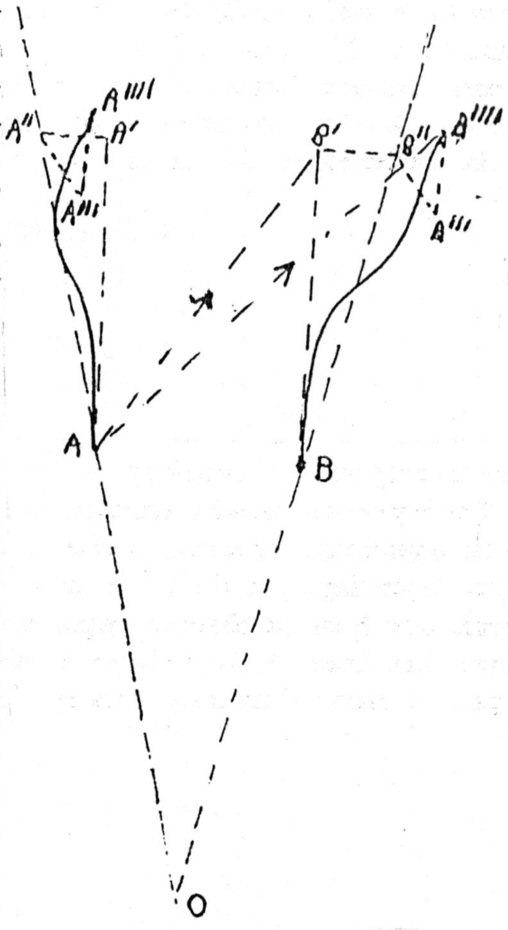

The conic path of light from A to B"" also resembles the form of a Maxwellian wave, having both lateral displacement and rotating polarity.

The hyperconic formulation introduces a different probability wave from that of the Schroedinger equation, in which two fixed boundaries are set for the wave group in one dimension, albeit set at infinity, but capable of producing a standing probability wave by reflection, like the plucked string of a violin. The peak of the probability wave then occurs near the centre of the string and goes to zero at infinity in either direction.

APPENDIX EIGHT

THE MATHEMATICS OF A HYPERCONE UNIVERSE

The systems approach to cosmology may appear novel, but we do have observable systems in the universe – the solar system and the galactic system, which behave in accordance with the dynamics of linear control systems in engineering design. It is therefore not such a great step to extend the criteria for these observed entities to the behaviour of the universe as a whole.

Since this type of universe is deemed to have a high translatory velocity, relative to its origin at the site of the big bang, it is appropriate to treat it as a combination of state-space trajectories in accordance with Liapunov's second method. This method can be applied equally to trajectory systems operating under restraint – like the water from a bath going down the plughole, where the plughole is a constraint, or to an aeroplane in free flight, which goes out of control and falls in a spin under the constraint of air resistance. Likewise to artillery shells at the limit of their range, which tend to tumble due to air resistance, and to satellites in orbit under gravitational constraint, which must be prevented from tumbling by steering jets.

The state-space is defined as the n-dimensional space in which the components of the state vector represent its co-ordinate axes. The unforced response of a system released from any initial point x (t_o), traces a curve or trajectory in an n-dimensional state-space. Time, t, is an implicit function along the trajectory. When the state variables are represented by phase variables, the state space may also be called a phase space. While it is impossible to visualise a space with more than three dimensions, the concept of a state space is nonetheless very useful in systems analysis.

The behaviour of second order systems is conveniently viewed in the state space, since it is two-dimensional and is, in fact a state (or phase) plane. It is easy to obtain a graphical or geometrical interpretation of second-order system behaviour in the state plane. The concepts developed for the state-plane can then be extrapolated to the state-space of higher-order systems.

The symmetry that results in the state-plane trajectories can be demonstrated by a matrix converted to normal, canonical form. Since there is no forcing function the response results from the energy stored in the system as represented by the initial conditions. The state equation is therefore:

$$x = \begin{vmatrix} a_{11} & a_{12} \\ a_{21} & a_{22} \end{vmatrix} x$$

The second-order example which is appropriate, is that one where the eigenvalues are complex with negative real parts, i.e., an underdamped system. This is because the development of the system has no constraints.

The equation representing the slope of the trajectory in the w_1, w_2 plane is:

$$dr/dq = -\frac{\sigma}{wd}\, r$$

The variables can be separated and the integration of the equation yields:

$$r = Ce^{-(\sigma/wd)\phi} \qquad = Ce^{(\zeta/\sqrt{1-\zeta^2})\phi}$$

For any initial values at time t = 0 of w_1 (0) and w_2 (0). The value of C is then given by:

$$C = r_o/e^{(\zeta/\sqrt{1-\zeta^2})\phi}$$

The trajectories represented by this equation are logarithmic spirals. With a negative damping ratio, $-1 < \zeta < 0$, the response is unstable and the trajectories diverge from the equilibrium point.

When a gravitating mass is introduced into the system as a form of constraint, the trajectories approach the equilibrium point, the stable focus of the mass, in a decreasing logarithmic spiral. This would explain why the orbits of the planets circling the sun can all be placed along a logarithmic spiral.

When this is extrapolated into a third order system the spiral is extended around the eigenvector in the x plane like a coil spring.

Reference: *Linear Control System Analysis and Design* – John J. D'Azzo & Constantine H. Houpis – McGraw-Hill, 1988, p. 517.

TABLES OF QUARK CHARACTERISTICS

Table One – Quarks and their Quantum Numbers:

Symmetry matrix group — — — SU (2) — — —
of particles — — — — — — — SU (3) — — — — — —
— — — — — — — — — — — SU (4) — — — — — — —

Flavour	Up	Down	Strange	Colour
Charge	2/3	–1/3	–1/3	2/3
Isospin	1/2	1/2	0	0
Isospin space	+1/2	–1/2	0	0
Strangeness	0	0	–1	0
Charm	0	0	0	1

Isospin is in units of $h/4\pi$; Isospin space is defined by electromagnetism; Strangeness arises in difference of decay modes; Charm is required by Pauli exclusion principle.

Particles in the SU (2) group

Particle	Proton	Neutron	Λ^{o}	Σ^{+}	Σ^{o}	Σ^{-}	Ξ^{o}	Ξ^{-}
Quarks	uud	udd	uds	uus	uds	dds	uss	dss

Particles in the SU (3) group

Particle	Δ^{++}	Δ^{+}	Δ^{o}	Δ^{-}	Σ^{+*}	Σ^{o*}	Σ^{-*}	Ξ^{o*}	Ξ^{-*}	Ω^{-}
Quarks	uuu	uud	udd	ddd	uus	uds	dds	uss	dss	sss

In order to create charmed hadrons a fourth flavour of charmed quark is required, and these four flavours form the basic representation of SU (4).

Particle	Σ_c^{++}	Σ_c^{+}	Σ_c^{o}	S^{+}	S^{o}	T^{o}
Quarks	cuu	c(ud+du)	cdd	c(us+su)	c(ds+sd)	css

(*An Introduction to Quarks and Partons*, F.E. Close, Academic Press, 1979 and *The Quest for Quarks*, Brian McCusker, Cambridge, 1983.)

BIBLIOGRAPHY

Page

3 *The Feynman Lectures on Physics,* Vol. 1, Chap. 8.

4 Rivolo A.R. *Nature,* Vol. 317, p. 788 (31.10.85), 'Velocity of Galaxies'.

5 *Guinan & Maloney, *New Scientist,* p. 23 (29.8.85), 'Double Star System Violates Relativity'.

" †Yahil A., Tamman G.A. & Sandage A. *Astrophysical Journal,* 217, 903 (1977).

" ††Hecht J. *New Scientist,* p. 25 (13.2.86). 'Is the Universe made of Froth?'

7 Combley F. *Physics Review,* Letters, Vol. 42, p. 1383 (5/79), 'Test of Special Relativity'.

" Cooper P.S. *Physics Review,* Letters, Vol. 42, p. 1386 (5/79), 'Test of Special Relativity'.

13 Brillet A., Hall J.L. *Physics Review,* Letters, Vol. 42, pp. 549-52 (2/79), 'Isotropy of Space'.

17 Reasenberg R.D. *Astrophysics Joint Letters,* (6/79). Cambridge, Mass., 'Time Delay'.

18 *Mather John: *New Scientist,* p. 48 (16.1.86), 'A Look at the Primeval Explosion by COBE'.

" †Bond J.R., Bergh S. *Nature,* Vol. 320, p. 489 (10.4.86), 'Deviations from Universal Expansion'.

19 *Meiksin A., Davies M. *Astronomical Journal,* Vol. 91, p. 191, 'Anisotropy in CBR Background'.

" †Turner E. *New Scientist,* p. 28 (15.5.86), 'Anisotropy of Space'.

" ††Robertson N.A. *Nature,* Vol. 320, p. 506 (10.4.86), 'Anisotropy of Space'.

20 Narliker J. *Violent Phenomena in the Universe,* Oxford University Press, 1982.

21 Linde Adrei: *Reports on Progress in Physics,* Vol. 47, p. 925. Libender Institute, Moscow.

23 Laurie John: 'Cosmology Now', BBC, 1973.
24 Weinberg Steve: *The First Three Minutes,* Basic Books, 1977.
27 Werner S.A., Colella R., Overhauser A.W., Eagen C.F. *Physics Review,* Letters, Vol. 35, p. 1053 (10/75), 'Phase Shift of Neutron in Magnetic Field'.
35 Bondi H. *Cosmology,* C.U.P., 1968.
36 Satz H. *Nature,* Vol. 324, p. 116 (13.11.86), 'The Quark Plasma'.
37 Ginzberg V.L., Sernoi L.M. *Astrophysics and Space Science,* (Netherlands) Vol. 48, No. 2, pp. 401-20 (5/77).
38 McCrea W.H. *Astrophysical Journal,* Vol. 98, p. 52 (4/78), 'Variable G'.
43 Rosner R., Weiss N.O. *Nature,* Vol. 317 (31.10.85), 'Differential Rotation and Magnetic Torque in Interior of Sun'.
47 Woltjer J. *Columbia University Scripta Varia,* 35 (1971), 'Massive Rotators in Galactic Nuclei'.
56 Shellard E.P.S., Brandenberger R.H., Kaiser N., Turk N. *Nature,* Vol. 326, p. 672 (16.4.87), 'Large Scale Velocities in Cosmic Strings'.
84 Ejsberg: *Fundamentals of Modern Physics,* p. 692. Wiley.
107 *Herbert Nick: *Quantum Reality.* Rider, 1986.
" †Rae Alastair: *New Scientist,* p. 36 (17.11.86), 'ESP and Quantum Physics'.
109 Smolin Lee: *New Scientist,* p. 40 (14.10.85). 'What is Quantum Mechanics About?'
121 *Birrel N.D., Davies P.C.W. *Journal of Physics,* A 13, 2104, 1980a. *Physics Review,* d22, 322, 1980b.
" †Prof. Hawkinge S.W. (Comm. Math.): *Physics,* 43, 199, 1975, et al.
133 Solymar L. *Lectures on E.M. Theory,* p. 47. Oxford Science, 1984.
135 Maddox J. *Nature,* Vol. 323, p. 199 (18.9.86), 'Helices with Optical Aativity'.
147 Officer C.B., Hallam A., Drake C.L., Devine J.D. *Nature,* Vol. 326, p. 143 (12.3.87), 'Faunal Extinctions'.
154 Ganguly S. *Nature,* Vol. 320, p. 511 (10.4.86), 'ULF Waves in the Ionosphere'.
157 Alfven Hannes, Falthammar Carl Gunne, *Cosmical Electrodynamics,* p. 14. Clarendon, 1963.

159 Hogan C.J., White S.D.M. *Nature,* Vol. 321, p. 575 (5.6.86), 'Galaxy Formation by Mock Gravity'.
179 Henley E.M., Thirring W. *Elementary Quantum Field Theory,* p. 28 McGraw-Hill.
193 Calder Nigel: 'The Key to the Universe', BBC, 1977.